GOLDEN EAGLES

LEGENDARY BIRDS OF PREY

BENTELI

HEINZ M. GUT

EX HIS QUAS
NOVIMUS AQUILAE
MAXIMUS HONOS,
MAXIMA ET VIS.

Of all the birds with which we are acquainted, the eagle is looked upon as the most noble, and the most remarkable for its strength. \01\

GAIUS PLINIUS SECUNDUS \
CA. 23—79

PREFACE

After years of work in the Swiss Alps, it is a great pleasure to see this book about golden eagles, which had existed for so long only in my thoughts, materialize and become a reality. On innumerable days in snow and ice or scorching heat, filled with blood, sweat and tears, I carried my heavy backpack up steep slopes, ravines, and crags, and waited for hours on exposed and windy cliffs in the hope of photographing the mighty king or queen of all birds. Every time the majestic bird of prey flaunted its incredible flight skills, waves of extreme excitement made my heart race and my soul rejoice. Every time the fearsome raptor could not be found, frustration and emptiness filled my body. On such days, I found comfort in admiring the marvelous mountain scenery and telling myself: "Although I haven't spotted a golden eagle, I could spend another wonderful day in the secluded mountain wilderness. What a joy! So, let's try again! Maybe next time I will have more luck."

I feel immensely privileged and grateful to have spent many, many days with golden eagles while capturing their wonderful life on camera. The proud birds not only taught me a lot about their secret lives, although many mysteries remain, but also instructed me about the grandiose nature of which they are a part, while giving me valuable insight about myself. Thanks to them, I now know who I am, where I belong, what life I would like and where my path will lead. Sometimes, the strength of my connection with these majestic creatures convinces me that I once was a golden eagle in an earlier life. In these moments, I recall the faint memories of when I jumped from a tall cliff and spread my large wings to soar along sunlit, vertiginous rock walls in the infinite space of the mountains, under a perfectly blue sky.

Golden eagles have fascinated mankind for centuries due to their proud and fierce appearance, impressive size, unmatched flight skills, bold hunting tactics, and noble character. In their wild and free lives, untethered from the earth and embodying freedom and supernatural powers, they naturally became potent idols that have been admired, feared, and worshipped throughout history in the most diverse cultures around the globe. Astonishingly, these mighty birds have lost nothing of their fascination and immediately conquer the heart of the keen observer, even today. Golden eagles have a deeply mystic presence that has the potential to change your life. Suddenly, after an encounter with a golden eagle, you might ask yourself what you really want from life, and whether it is time to change and live closer to nature, becoming more free, grounded, and fulfilled.

This book aims to pay tribute to the golden eagle in the hope of opening hearts, not only for the king of birds, but also for all the magnificent creatures on this planet and for nature in general. If we really open our hearts, minds, and eyes, we will recognize how incredible life, nature, and our planet really are. As a result, we will never forget where we come from and always be ready to live with nature, not in spite of it.

Heinz M. Gut

INTRODUCTION

06 — 23

THE LIFE OF GOLDEN EAGLES

24 — 40

IN FAIRY TALES, MYTHS, POEMS, AND QUOTES

41 — 184

MOUNTAIN WILDERNESS: ENCOUNTERS, ADVENTURES, AND DIFFICULTIES

185 — 218

OUTRO

219 — 224

How golden eagles conquered my heart

Back from another wildlife photography trip in Africa, I wondered about opportunities to photograph wild animals in Switzerland. Although neither elephants, rhinos, hippos, giraffes, nor lions and cheetahs roam my home country, the rugged mountains of Switzerland are full of fascinating and beautiful wild creatures. My plan was to document the life of an impressive predator of Switzerland, a choice which narrowed the possibilities to a handful of species. Wolf, bear, lynx, and the golden eagle are the classic apex predators of the Alps and each of these magnificent beasts would have been an equally wonderful species for a photographic journey. After more than hundred years of absence, the top predators, wolf and bear, were just coming back into the country, creating jubilation among wildlife enthusiasts while triggering an angry uproar from most farmers and hunters. Despite the huge temptation to look for these new magnificent arrivals, the numbers of these terrific predators were still extremely low when I initiated the project, with only two or three bears and a dozen wolves that had made their way from Italy into Switzerland, hoping for new territories. The lynx on the other hand, although present in much higher numbers, is a nocturnal predator that is infamous for its skill in avoiding any interaction with humans. At the time, I concluded that these large predators would therefore be very difficult to find and study with the camera. But when I thought about the golden eagle, faint memories from my childhood and later life as a rock climber came back to me. While on a hike or mountaineering with my father as a small boy in the Alps, I remember him suddenly pointing up high to a peak or into the sky and say: "Look, there flies a golden eagle. He is the king of the sky." Although I could only spot the small silhouette of a large bird that soared alongside terrifying rock faces, glided elegantly around snow-white peaks, or cruised through the infinity of the blue sky, I looked at the magnificent eagle with great respect and thought: "I feel so privileged and proud to see such an amazing creature." Later on, in my life as a rock climber, I recollect moments ascending a difficult wall, when suddenly a large shadow passed rapidly over me and along the rock face. Curious what might have covered the sun for an instant, I turned my head and recognized an immense bird high above me, gliding effortlessly along the cliff. The airiness of the large bird's skillful flight was in such contrast to my situation where the earth's mass mercilessly pulled on my body. "A golden eagle! What an incredible bird high up in the sky, for whom gravity seems not to exist. What would I give to lead such a life one day?" I thought in these moments, and continued my struggle against gravity up the route to the next anchor.

But flies an eagle flight,
bold, and forth on,
leaving no track behind.

WILLIAM SHAKESPEARE \
1564 — 1616

Encouraged by such memories and my life-long passion for the mountains with their cliffs, peaks, glaciers, and canyons providing copious alpine wildlife habitats, the choice was quickly made and the Golden Eagle Project was born. "This is the predator species in Switzerland that I want to study and follow with the camera." I decided. Golden eagles are majestic birds of prey and a phenomenal object for a wildlife documentary. But with only 350–360 breeding pairs in Switzerland and territories as large as 200 km^2, they are notoriously difficult to photograph.

Only after years of learning and dozens of excursions into the mountains, was I able to find locations high up in rock walls from where I started to capture the life of the wild predator birds on camera. What had started as a small project developed over the years into a great passion that demanded more and more time. I understood that high-quality photographs of golden eagles have to be earned by persistence, a strong commitment, and absolute dedication. Consequently, I needed to spend as much time as possible in the mountains. To pass days on steep rock faces and observe huge golden eagles fly along sunbathed cliffs is indeed an incredible privilege.

Despite their majesty, golden eagles lead a harsh life where every day is a battle for survival. These grand raptors withstand temperatures of -20° in winter as well as heat waves in the summer months. They endure weeks without food and constantly need to defend their home range against other individuals who would like to install themselves in the same place. The daily quest in search of prey or patrolling their territory has turned them into matchless experts of thermals, effortlessly traversing vast distances over alpine valleys and peaks. Without any doubt, golden eagles have become the mighty kings and queens of the mountain sky.

Due to their incredible skills and powerful appearance, these magnificent and legendary birds of prey have fascinated and inspired humans throughout history, over continents, and in the most diverse cultures and beliefs. Admired, worshipped, and feared; eagles have been used as mighty symbols of freedom, beauty, and strength, but also as symbols of terror and fear since ancient times.

08 Taxonomy and systematics of eagles

Although the present work focuses on the majestic golden eagle, it is important first to get an overview of the large group of birds of prey that are commonly referred to as eagles. Eagles can be found over almost the entire globe, and in many places these mighty raptors have become spectacular icons of the local fauna. The eagle's large and powerful appearance with a heavy head, large hooked beak, strong muscular legs with sharp talons, a piercing stare out of colorful eyes, and the often very impressive body size and large wingspan have made these birds of prey one of the most iconic animal species on the planet. Since eagles are usually much larger than any other raptor, except for some vultures, they are placed at the top of the food chain as apex predators. The term eagle is used for dozens of birds of prey that belong to the *Accipitridae*, an avian family that encompasses around 14 subfamilies, 65 genera, and 231 species.[02]

Based on classical taxonomy and genetic analyses, eagles have been classified into four *Accipitridae* subfamilies that are further divided into several genera. The *Circaetinae* subfamily describes snake eagles containing Old World species that mostly consume snakes, reptiles, or small mammals in their diet. To protect themselves from snake bites, they usually have heavy scales on their legs and, as an adaptation to their relatively small prey, they have short toes. The *Circaetinae* subfamily contains six genera, of which four are monotypic meaning they only contain one species.[03]

→ TABLE 1 \ PAGE 10

Next is the *Aquilinae* subfamily that contains booted eagles and hawk eagles which are distributed all around the world, where they populate highly diverse habitats. In general, these eagles are large raptors with feathered tarsi which target a wide variety of prey. Five out of the ten genera are monotypic. The *Aquilinae* is the largest subfamily with 38 different species. The golden eagle has its place in the *Aquilinae* subfamily.[03]

→ TABLE 2 \ PAGE 11 \ 12

Then comes the *Haliaeetinae* subfamily, which encompasses sea and fish eagles from all over the world that prefer rivers, lakes, and coastal regions of the oceans as habitat. A large part of their diet is fish, either as fresh catch or rotten carcass. Taxonomic features include bare tarsi and a fused second joint of the inner toe. All *Haliaeetinae* species are now consolidated into one genus.[03]

→ TABLE 3 \ PAGE 13

The last subfamily is the *Harpiinae*, containing large and powerful raptors of the tropical forest. These birds of prey feed on medium-sized mammals like monkeys or sloths that live on the towering trees of the jungle. Members of the *Harpiinae* also have unfeathered legs. All three genera of the *Harpiinae* subfamily are monotypic.[03]

→ TABLE 4 \ PAGE 13

Interestingly, eagle species are not evenly distributed over the planet. Eurasia and Africa have the highest diversity, featuring more than 60 different species; while in the entire rest of the world only 14 species are found: two in North America, nine in Central and South America, and three in Australia.

Hunting strategies and prey vary greatly from species to species and are closely related to habitat specialization as well as abundance and type of prey. Sea eagles from the *Haliaeetus* genus, for example the African fish eagle, are highly skilled in capturing fish from the ocean, in freshwater rivers, and in lakes; but they also hunt amphibians and water birds. Raptors that belong to the *Circaetinae* subfamily, like the black-chested snake eagle which is found throughout Southern and Eastern Africa, predominantly hunt reptiles like snakes, serpents, or lizards. Even though snake or serpent eagles have adapted to very different habitats, ranging from open savannahs to dense tropical forests, they all have specialized in tracking and killing all kinds of reptiles. Imperial eagles, golden eagles, tawny eagles, and other members of the genus *Aquila* are apex predators in open or semi-open habitats, where they prey on small to medium-sized vertebrates from squirrels and hares to foxes and ungulates such as chamois or small antelopes. Other eagle species occupy highly complex niches and therefore have developed unique anatomical features to hunt animals that are typically present in their habitat. One example is the powerful and impressive harpy eagle, the sole member of the genus *Harpia*, which possesses the largest talons of all *Accipitridae*, enabling it to hunt sloths and monkeys high up in giant trees of the South American tropical forest and carry them back into the nest for feeding the young.[04] Another example of extreme specialization is Verreaux's eagle, a large black eagle found in Southern and Eastern Africa, which hunts and feeds almost exclusively on rock or cape hyrax.[05, 06] In this case, prey and predator share a highly specific habitat composed of small rocky hills in an arid environment, often covered with savannah or thorn bush vegetation, which can be found in rocky hills and mountains, cliffs along ephemeral rivers, or the inselbergs of the Namib Desert.

Many eagles are ambush hunters, surprising their prey with attacks from unexpected angles, through dense forest, or by covering themselves behind topographical features like hills, rocks, or bushes, or in low ground flights, before they attack. Prey are often snatched in flight without landing and taken back to a nest or perch to feed on. Large eagles may kill prey that is much heavier and larger than the raptors themselves. Golden eagles, or martial eagles, are well known to hunt large ungulates (chamois, doe, young deer, and small antelopes) with body weights well over 30 kg. In this case, carcasses are devoured on the site of the kill or dissected into parts which are carried back to a nest or perch.

Eagle nests are called eyries and are built either in old and robust trees or on rock ledges high up in inaccessible cliffs. Many species lay two or sometimes three eggs from which chicks hatch and quickly develop. The first hatchling often grows much faster, becoming stronger and taller than its younger siblings, who can significantly lag behind in their development. The sturdiest chick frequently snatches much more food from the feeding parents compared with the weaker siblings, who struggle getting enough nourishment and become weaker and weaker. Frequently the dominant chick, often the first born, kills its underdeveloped siblings without intervention by the parents. This is a natural behavior known as cainism, and is common for several eagle species, for example Verreaux's eagle or tawny eagle, while it happens only under stress conditions such as food shortage in other eagle species (e.g. golden eagle, imperial eagle).[07]

The size and body weight of eagles varies greatly from smallish birds, such as the pygmy eagle found in New Guinea or the little eagle from Australia that have body sizes of about 0.5 m and wingspans of about 1.2 m, to some of the largest birds known on the planet with body sizes up to 1 m, body mass exceeding 6 kg, and wingspans of more than 2 m. Among these large and impressive raptors are the Philippine eagle, Steller's sea eagle, the white-tailed sea eagle, and the harpy eagle, martial eagle, and the golden eagle among others.
In the order of the *Falconiformes*, these huge birds of prey are only surpassed by most Old World vultures, for example the enormous lapped-faced vulture of sub-Saharan Africa, and two species of New World vultures, the Californian condor and the Andean condor.

TABLE 1

Eagle species that belong to the *Circaetinae* subfamily of the *Accipitridae*

COMMON NAME	SCIENTIFIC NAME	GENUS	DISTRIBUTION
ANDAMAN SERPENT EAGLE	*Spilornis elgini*	*Spilornis*	Endemic to the Andaman Islands, India
PHILIPPINE SERPENT EAGLE	*Spilornis holospilus*	*Spilornis*	Endemic to the Philippines
MOUNTAIN SERPENT EAGLE	*Spilornis kinabaluensis*	*Spilornis*	Endemic to Borneo
SULAWESI SERPENT EAGLE	*Spilornis rufipectus*	*Spilornis*	Endemic to Sulawesi, Indonesia
NICOBAR SERPENT EAGLE	*Spilornis klossi*	*Spilornis*	Endemic to Nicobar Islands, India
CRESTED SERPENT EAGLE	*Spilornis cheela*	*Spilornis*	Indian subcontinent, Southeast Asia, Indonesia, Borneo
BANDED SNAKE EAGLE	*Circaetus cinerascens*	*Circaetus*	Tropical Africa
FASCIATED SNAKE EAGLE	*Circaetus fasciolatus*	*Circaetus*	Tropical East Africa
BROWN SNAKE EAGLE	*Circaetus cinereus*	*Circaetus*	Sub-Saharan Africa
BLACK-CHESTED SNAKE EAGLE	*Circaetus pectoralis*	*Circaetus*	East and Southern Africa
BEAUDOUIN'S SNAKE EAGLE	*Circaetus beaudouini*	*Circaetus*	Sahel, West Africa
SHORT-TOED SNAKE EAGLE	*Circaetus gallicus*	*Circaetus*	Mediterranean area, Western Russia, Middle East, Indian subcontinent, Western Asia, Indonesia
BATELEUR	*Terathopius ecaudatus*	*Terathopius*	Sub-Saharan Africa
MADAGASCAR SERPENT EAGLE	*Eutriorchis astur*	*Eutriorchis*	Endemic to Madagascar
PHILIPPINE EAGLE	*Pithecophaga jefferyi*	*Pithecophaga*	Endemic to the Philippines
CONGO SERPENT EAGLE	*Dryotriorchis spectabilis*	*Dryotriorchis*	African tropical rainforest, lower Guinean forests

TABLE 2

Eagle species of the subfamily *Aquilinae*

COMMON NAME	SCIENTIFIC NAME	GENUS	DISTRIBUTION
GOLDEN EAGLE	*Aquila chrysaetos*	*Aquila*	North America, Eurasia, North Africa
VERREAUX'S EAGLE	*Aquila verreauxii*	*Aquila*	Central and East Africa
TAWNY EAGLE	*Aquila rapax*	*Aquila*	Northern, Central, and Southern Africa, Indian subcontinent, Arabian Peninsula
STEPPE EAGLE	*Aquila nipalensis*	*Aquila*	Arabian Peninsula, East and Southern Africa, Indian subcontinent, Central Asia
SPANISH EAGLE	*Aquila adalberti*	*Aquila*	Spain, Portugal
IMPERIAL EAGLE	*Aquila heliaca*	*Aquila*	East-central and Southeastern Europe, Arabian Peninsula, Northeast Africa, Northern India, Pakistan, Central and East Asia
GURNEY'S EAGLE	*Aquila gurneyi*	*Aquila*	Indonesia, New Guinea
WEDGE-TAILED EAGLE	*Aquila audax*	*Aquila*	Endemic to Australia
BONELLI'S EAGLE	*Aquila fasciata*	*Aquila*	Mediterranean area, Indian subcontinent, Arabian Peninsula, China
AFRICAN HAWK-EAGLE	*Aquila spilogaster*	*Aquila*	Western, Central, Southern, and East Africa
CASSIN'S HAWK-EAGLE	*Aquila africana*	*Aquila*	Western and Central Africa
LEGGE'S HAWK-EAGLE	*Nisaetus kelaarti*	*Nisaetus*	Indian subcontinent and Sri Lanka
MOUNTAIN HAWK-EAGLE	*Nisaetus nipalensis*	*Nisaetus*	North India, Nepal, Pakistan, Southeast Asia, China, Taiwan, Japan
WALLACE'S HAWK-EAGLE	*Nisaetus nanus*	*Nisaetus*	Malaysia, Sumatra, Borneo
JAVAN HAWK-EAGLE	*Nisaetus bartelsi*	*Nisaetus*	Endemic to Indonesia
SULAWESI HAWK-EAGLE	*Nisaetus lanceolatus*	*Nisaetus*	Endemic to Indonesia
BLYTH'S HAWK-EAGLE	*Nisaetus alboniger*	*Nisaetus*	Malaysia, Singapore, Sumatra, Borneo
PINSKER'S HAWK-EAGLE	*Nisaetus pinskeri*	*Nisaetus*	Endemic to the Philippines
PHILIPPINE HAWK-EAGLE	*Spizaetus melanoleucus*	*Nisaetus*	Endemic to the Philippines
FLORES HAWK-EAGLE	*Spizaetus tyrannus*	*Nisaetus*	Endemic to Indonesia
CHANGEABLE HAWK-EAGLE	*Spizaetus isidori*	*Nisaetus*	Indian subcontinent, Southeast Asia, Indonesia, Borneo, Philippines

TABLE 2

Eagle species of the subfamily *Aquilinae*

COMMON NAME	SCIENTIFIC NAME	GENUS	DISTRIBUTION
BLACK-AND-WHITE HAWK-EAGLE	*Nisaetus philippensis*	*Spizaetus*	Mexico, Central America, South America
BLACK HAWK-EAGLE	*Nisaetus floris*	*Spizaetus*	Mexico, Central America, South America
BLACK-AND-CHESTNUT EAGLE	*Nisaetus cirrhatus*	*Spizaetus*	South America
ORNATE HAWK-EAGLE	*Spizaetus ornatus*	*Spizaetus*	Mexico, Central America, South America
BOOTED EAGLE	*Hieraaetus pennatus*	*Hieraaetus*	Southern Europe, North Africa, Arabian Peninsula, Middle East, sub-Saharan Africa, Southern Africa, Indian subcontinent, Central Asia
WAHLBERG'S EAGLE	*Hieraaetus wahlbergi*	*Hieraaetus*	Sub-Saharan Africa
LITTLE EAGLE	*Hieraaetus morphnoides*	*Hieraaetus*	Endemic to Australia
PYGMY EAGLE	*Hieraaetus weiskei*	*Hieraaetus*	Endemic to New Guinea and Moluccas
AYRES'S HAWK-EAGLE	*Hieraaetus ayresii*	*Hieraaetus*	Sub-Saharan Africa
INDIAN SPOTTED EAGLE	*Clanga hastata*	*Clanga*	Indian subcontinent
LESSER SPOTTED EAGLE	*Clanga pomarina*	*Clanga*	Eastern Europe, East Africa
GREATER SPOTTED EAGLE	*Clanga clanga*	*Clanga*	Central and Western Asia, Middle East, Arabian Peninsula, Indian subcontinent, Southeast Asia
LONG-CRESTED EAGLE	*Lophaetus occipitalis*	*Lophaetus*	Sub-Saharan Africa
BLACK EAGLE	*Ictinaetus malaiensis*	*Ictinaetus*	Indian subcontinent, Southeast Asia, Indonesia, Borneo, China
MARTIAL EAGLE	*Polemaetus bellicosus*	*Polemaetus*	Sub-Saharan Africa
RUFOUS-BELLIED EAGLE	*Lophotriorchis kienerii*	*Lophotriorchis*	India, Sri Lanka, Southeast Asia, Indonesia, Philippines, Borneo
CROWNED EAGLE	*Stephanoaetus coronatus*	*Stephanoaetus*	Sub-Saharan Africa

TABLE 3

Eagle species of the subfamily *Haliaeetinae*

COMMON NAME	SCIENTIFIC NAME	GENUS	DISTRIBUTION
WHITE-BELLIED SEA EAGLE	*Haliaeetus leucogaster*	*Haliaeetus*	India, Sri Lanka, Southeast Asia, China, Australia
SANFORD'S SEA EAGLE	*Haliaeetus sanfordi*	*Haliaeetus*	Endemic to Solomon Islands
AFRICAN FISH EAGLE	*Haliaeetus vocifer*	*Haliaeetus*	Sub-Saharan Africa
MADAGASCAR FISH EAGLE	*Haliaeetus vociferoides*	*Haliaeetus*	Endemic to Madagascar
PALLA'S FISH EAGLE	*Haliaeetus leucoryphus*	*Haliaeetus*	Central Asia, China, Pakistan, Nepal, Northern India, Bangladesh, Myanmar
WHITE-TAILED EAGLE	*Haliaeetus albicilla*	*Haliaeetus*	Western Greenland, Eastern Europe, entire Eurasia to Kamchatka
BALD EAGLE	*Haliaeetus leucocephalus*	*Haliaeetus*	North American continent (Alaska to Mexico)
STELLER'S SEA EAGLE	*Haliaeetus pelagicus*	*Haliaeetus*	Coasts of the Japan and Okhotsk Seas
LESSER FISH EAGLE	*Haliaeetus humilis*	*Haliaeetus*	Kashmir through Southeast India, Nepal and Burma towards Indochina
GRAY-HEADED FISH EAGLE	*Haliaeetus ichthyaetus*	*Haliaeetus*	India, Southeast Asia

TABLE 4

Eagle species that belong to the *Harpiinae* subfamily

COMMON NAME	SCIENTIFIC NAME	GENUS	DISTRIBUTION
HARPY EAGLE	*Harpia harpyja*	*Harpia*	Tropical Central and South America
NEW GUINEA EAGLE	*Harpyopsis novaeguineae*	*Harpyopsis*	New Guinea
CRESTED EAGLE	*Morphnus guianensis*	*Morphnus*	Tropical Central and South America

Eagles as mighty symbols in human culture and mythology

Mankind has admired eagles for millennia because of their impressive hunting capacities, unmatched eyesight, incredible flight skills, and puissant look. It is therefore no surprise that eagles already made their way into religious and cultural practices of humans very long ago. The mighty birds of prey are omnipresent in tales, legends, oral accounts, and origin myths of innumerable cultures all around the world, where they appear in a surprisingly diverse sociocultural context. In these varying cultural customs, the eagle itself, or even just its powerful beak or fearsome talons, represents the divine, the sky, the sun, lightning, a thunderbolt, or sunrays. Hence, eagles are often attributes of gods, giants, or other mythical creatures that account for some of their supernatural powers, or are used to make sense of inexplicable natural phenomena such as storms, earthquakes, lightning or thunder. The modern English term eagle for the name of the largest birds of prey comes from the Latin word *aquila* (eagle) which became aigle in old French and egle in old English.

In Indian mythology, the eagle is omnipresent in the form of Garuda, an immortal bird-god who serves as the vahana, the deity's vehicle, of god Vishnu.[08] Garuda is described as the king of all birds and appears as a giant eagle in its zoomorphic form or as half human, half eagle, in its anthropomorphic form. Garuda is a mighty and feared warrior with heavenly powers that symbolizes speed, violence, and military prowess. Although already present in the earliest Hindu Vedas, dating to around 1500 BC, the bird's heroic appearances in the Mahabharata, along with the Ramayana, the major Sanskrit epic of ancient India, made him very popular among believers. Garuda stands for the divine king of all birds who possesses absolute courage, is fast, can assume any form, enter anywhere, and can even stop the spinning of the earth, heaven, and hell, just by flapping its giant wings.[08, 09] Garuda is often present in paintings, reliefs, or sculptures of Vishnu or his avatars, either as the Lord's mount or next to him.[10] Garuda is also the prime enemy of snakes and supports Lord Vishnu, the preserver and protector of the universe, with its incredible force and prowess in his regency.

> We were free as the winds,
> and like the eagle,
> heard no man's commands.
>
> RED CLOUD (MAȞPÍYA LÚTA IN LAKOTA) \
> 1822—1909

With the spread of Indian mythology over the Indian subcontinent and into Central, Southeast, and East Asia, and its evolution into classic Vedic traditions such as Hinduism, Buddhism, or Jainism, Garuda became ubiquitous over the entire Asian continent. From the earliest Buddhist monuments of Gandhara (today's Northwest Pakistan and Northeast Afghanistan), across the Angkor capital of the Khmer empire in Cambodia, to Buddhist monasteries of Lhasa in former Tibet, and temples in Myanmar, Nepal, Mongolia, Indonesia, Philippines, Thailand, or Japan, variations of Garuda are found everywhere. Even in modern times, Garuda is not only present in Hindu-Buddhism-Jainism religious contexts all over Asia but even became the national emblem of Indonesia and Thailand. In Indonesia, Garuda Pancasila is the emblem of the official government within its ministries, departments and institutions. It is depicted on palaces, monuments, government offices, and embassies abroad, but also in classrooms next to the president's photograph, or on uniforms and flags of military and police. Similarly, a depiction of

Garuda was used for centuries by ancient kings of Thailand before it became the official emblem of the modern Thai government. It is present on official government buildings, police and military uniforms, and is used to verify official documents such as passports, bank notes, or court orders.

In the Norse mythology of Scandinavia, dating to pre-Christian times, an eagle sits on Yggdrasil, the enormous ash tree in the center of the cosmos that connects the Nine Worlds.[11, 12] In constant enmity with the snake or dragon Nidhogg, who tries to bring Yggdrasil and with it the entire cosmos to fall by gnawing on one of the three giant roots, the eagle symbolizes foresight and knowledge as he has the overview over the world. Ratatosk, a squirrel that also lives on Yggdrasil, acts as a messenger to bring the eagle's words to Nidhogg in the underworld. Another appearance of the eagle in Norse mythology is the giant Hräsvelgr, who in the form of an eagle creates wind and storm with its giant wings.[13] In several other poems and chapters of the Poetic and the Prose Edda, the two medieval literary works of Norse mythology, other giants and gods, for example Odin, can also change into the form of an eagle to fly to other places.

The Eagle of Zeus (*aetos dios*) was a giant golden eagle that served as one of the attributes of Zeus, the god of the sky, earth, weather, law, victory, destiny, and kingship in ancient Greek mythology.[14] Even the king of all gods in the Olympian pantheon sometimes turned into an eagle, for example when he abducted the beautiful Ganymede from Troy, with whom he had fallen in love, to Mount Olympus where he became his personal cup-bearer.[15] Zeus' divine eagle guarded the sacred scepter and lightning bolt, and served as Zeus' messenger to spread the word about victory and the god's will to mortal humans. Like Zeus, *aetos dios* is a symbol for the divine and heavenly, as well as for light, fertility, and luck, but it also has a despotic side. After the Titan Prometheus had illegally stolen the fire to hand it over to humanity, Zeus sentenced him to eternal torment for his crime. The immortal Titan was tied to a rock in the mountains and each day an eagle, the attribute of Zeus, was sent to Prometheus to feed on his liver which then grew back the next night just to be eaten by the eagle again and again.[16]

When Ptolemy I Soter (367–282 BC), after Alexander the Great's death, became ruler of the Ptolemaic Kingdom in Egypt, he introduced the Hellenistic *aetos dios* in North Africa.[17] The golden eagle standing on a lightning bolt decorated Ptolemaic currency for almost 300 years until the end of the Ptolemaic Kingdom in 30 BC under its last ruler, Cleopatra.

In an attempt to characterize the natural world, Greek philosopher and polymath Aristotle (384–322 BC) described in his History of Animals (Book IX, part 34) the habit of the sea eagle to make his young stare directly at the sun. If a hatchling refused, he reports, the eagle used force and twisted its body into the harsh sunbeams. He describes how the eagle then killed the one whose eyes started to water because of the bright sunlight and raised the other.[18] Several other cultures later absorbed and integrated this myth into their beliefs, and the eagle staring directly into the sun became a symbol for the absolute Christian belief from the Early Middle Ages onwards.[19]

In the early Roman Empire, before the adoption of Christianity, an eagle was the sacred animal of the chief deity Jupiter, the god of the sky and thunder.[20] Similar to Zeus in ancient Greece, Jupiter was often depicted in the presence of a large golden eagle holding a thunderbolt in his talons. In the Roman Empire, the eagle became most famous as a powerful symbol of the Roman military in the form of *Aquila* on the standard of the Roman legion.[21]

Although the eagle had been introduced earlier as military insignia together with the wolf, ox (or Minotaur), horse, and boar, the *Aquila* standard became the sole military symbol for a legion after Gaius Marius' military reform in the year 104 BC.[22] A special legionary, the aquilifer, carried the eagle standard on a lance as an auspicious and religious symbol, and as identification insignia in front of a legion. Initially made of gilded silver, an eagle with wide outspread or lifted wings of pure silver or gold, holding a thunderbolt in his claws and the letters S.P.Q.R. (the Latin abbreviation for *Senatus Populusque Romanus*) written underneath, was used under Caesar and the subsequent emperors.[23, 24] To lose an *Aquila* standard to an enemy in battle was an ignominy that must be prevented at all costs. If, despite all measures, an *Aquila* standard got lost or was captured by the enemy, the Roman military undertook great efforts to get it back.

In the monumental work *Naturalis Historia*, Pliny the Elder dedicates in his discussions about zoology (books VIII to XI) a section to well-known eagle species of the time. He describes features and behavior of six different species, some of which are definitely eagles, while others might have been vultures. Another appearance of the eagle in the Roman Empire, that also greatly emphasizes its powerful symbolic meaning as divine animal, was the eagle's role in the process of apotheosis, whereby a deceased emperor was voted divine by the Senate, becoming god-like.[25] After a one week long mourning period, accompanied by religious celebrations and a parade through the Via Sacra in Rome, the body of a deceased emperor, or an image of it, was brought to Campus Martius where it was placed on a large funeral pile with a tall, tapering, multi-story, palace-like structure on it. Adorned with hangings full of gold thread, ivory sculptures, incense, fruits, and fragrances donated by noble men of the entire empire in honor of the emperor, a procession of horsemen and chariots was held around the pile. Lastly, the pile was set on fire and an eagle was let loose from the highest chamber of the construction to soar up to the sky in the ascending fire and carry the soul of the emperor from earth to heaven. After this ceremony, the emperor was worshipped like a god.[25]

It is only the eagle
that can give an account
of the whisper of the cloud.

AFRICAN PROVERB

After the Western Roman Empire came to an end in the year 476 AD, the Roman eagle reappeared as a symbol of power under Charlemagne of the Frankish Empire. After his coronation as emperor by Pope Leo III in Rome in the year 800, he became the first emperor to rule Western Europe since the fall of the Western Roman Empire 300 years before. From then on, the eagle was used as a symbol of the Carolingian Empire and its emperors, kings, and rulers throughout the Early and Late Middle Ages.[26, 27]

The eagle was also introduced as a sign of omnipotence in Christian symbolism. With its unsurpassed acuity of vision, its incredible strength, and the highest flight of all birds, the eagle was a perfect allegory for the almightiness and wisdom of god. The mighty bird is mentioned several times in the bible and stands for god's love, the power of believing, rejuvenation, and resurrection.[28] In the fifth century, the eagle became the symbol of evangelist John, who already in the prologue of his gospel was believed to have reached the highest spiritual awareness when talking about the Word and God, similar to an eagle soaring high in the sky and gazing straight into the light of the sun.[29] Habits of eagles described in ancient pagan fables also found their way into Christian beliefs and symbolism. One of these stories mentions an elderly eagle whose strength after a long life had started to weaken, and that would fly directly into the sun's corona to scorch its old feathers and to purify its eyes, before diving three times into a spring to reappear completely rejuvenated.[30, 31] Such tales were used to describe the renewal of youthful strength through faith in god. Similarly, the myth that eagles look directly into the sun when they take off reminded Christians to focus on the highest spiritual and eternal insight in their religion.[18, 19]

This is Garuda of great splendour
that bird capable of assuming at will any form,
of going at will everywhere,
and of calling to his aid
at will any measure of energy.

KRISHNA-DWAIPÂYANA VYASA \
3RD MILLENNIUM BC

Also on the American continent, eagles played a highly important role in many indigenous cultures as religious and spiritual symbols to represent the divine and sacred. Native American people of the United States and First Nations members in Canada have created eagle depictions in sculptures, paintings, and embroideries over centuries. In addition, insignia made of eagle parts were used in many ceremonies to serve as symbols of supernatural powers in the tribal communities.[32, 33] As eagles fly very high, Native Americans believed that they were able to see everything and therefore provide a connection to the divine creator. Due to the high flight, the unmatched vision, the courageous soul, the swift dives, and the powerful appearance, eagles were given the greatest respect. They stood for honesty, truth, wisdom, strength, courage, and freedom and are still used today in many tribal logos as mighty symbols of the Native American culture.[34]

In many Native American tribes such as the Zunis, Hopis, and Pueblos of Northern Mexico and of the southwest of the United States (today's Utah, Arizona, Colorado, and New Mexico), or the Comanche, Crow, and Blackfoot of the Great Plains and Canadian prairies, eagles were thought to be messengers between heaven and earth by carrying prayers to the Great Spirit or divine creator.[35] It was believed that they even control the weather or bring long awaited rain.[36] Many of the tribes all over the North American continent celebrate the supernatural powers of the golden or bald eagle by a spiritual "eagle dance" where dancers dress in costumes that look like eagles.[37] Arms are covered in large eagle feathers resembling wings while other feathers on the costume are for the bird's tail or its plumage-covered legs. With often richly adorned headdresses, dancers imitate the movements of a soaring, flapping, or hunting eagle, telling the story of its life from birth to death. The performance is accompanied by singing and drumming which make the eagle dance an impressive and highly spiritual celebration in honor of the king of all birds. The dance was often performed to seek help from the gods, or higher powers, to resolve issues like illness, drought, or birth and death, but also before battles or war.

In many Native American communities, hunting or killing of eagles to obtain the sought-after feathers was highly regulated and eating of eagle meat was often a taboo. In some tribes the killing of an eagle was forbidden, and in consequence fledglings were taken from the nest, raised by hand, and feathers then gently removed from juvenile or adult animals without causing harm. Extended families often united in clans and used animals for identification. Golden or bald eagles served as clan animals in many different communities such as the Hopis or Puebloans from the Southwest, the Osage from the Great Plains, or the Haida from the Northwest, where an eagle was often depicted on totem poles, staffs, and other ritual insignia.[38] The iconic ceremonial headgear worn by male leaders of the highest rank in many Native American tribes was very often made of, or decorated with, eagle feathers. The sharp contrast in the white-brown colored feathers of juvenile golden eagles resulted in the most spectacular war bonnets signifying the highest marks of respect that an honored person must have earned through wise action or outstanding courage.[39]

> Leaders are like eagles...
> they don't flock.
> You'll find them one at a time.
>
> KNUTE KENNETH ROCKNE \
> 1888—1931

The great importance of eagles in the Native American culture is also reflected in the so-called eagle feather law, which provides exceptions for bald and golden eagles in the Migratory Bird Treaty Act from 1918, the Endangered Species Act from 1973, and the Bald and Golden Eagle Protection Act. Among other points, it regulates the use of eagle parts, nests, or eggs for the religious purposes of Native Americans so that they can continue their cultural habits.[40] Members that belong to officially recognized tribes who can prove their Native American ancestry can lawfully obtain eagle feathers. Unauthorized individuals that collect, keep, or deal in eagles or parts thereof face heavy fines.

In ancient Mesoamerican cultures, such as the Aztecs or the Maya, eagles were associated with noble men, ritualistic sacrifice, sorcery, and warfare for centuries.[41] The kings of the sky were of exceptional significance in their beliefs and are therefore still present in modern societies of Central America. The largest eagles found in the habitats of the region are the harpy eagle, a fearsome predator of the tropical forest that preys on arboreal monkeys and sloths, and the golden eagle of Northern Mesoamerica that prefers open grassland and savannahs, where it hunts medium-sized ground-dwelling creatures like rabbits or young deer. Given the geographical distribution of their habitats, the harpy eagle was most present in the Mayan culture, while the golden eagle played a role in the religious art of the Mexica and other cultures of Northern Mesoamerica. The importance of the eagle in Mesoamerican traditions is also highlighted in the Tonalpohualli, the Aztec version of the 260-day almanac that dates back to early Olmec times and is used by most Mesoamerican cultures as a calendar that sets the dates of spiritual rituals by a combination of numbers (1 to 13) and 20 names. In this ancient calendar system the eagle (quauhtli) is used as one of the 20 day names.[42]

The importance of eagles in Mexico is exemplified in the Mexican flag, where an eagle perches on a cactus, feeding on a snake held in its talons. The picture originates in

Aztec traditions in which Huitzilopochtli, the god of war and the sun, told the tribes to move south and search for a new homeland.[41] He instructed them that they would have reached their new grounds as soon as they see an eagle seated on a cactus, devouring prey. The eagle on the cactus with a snake has become a powerful national symbol which defines Mexican identity and helps unite its people to this day.

The Aztecs were a highly developed culture of Central Mexico that flourished in the 14th and 15th centuries. In their traditions, the eagle was omnipresent as a symbol of their societal organization.[41] Both the jaguar, as the most powerful land animal, and the eagle as the strongest animal of the sky, were used as representations of earth and heaven. Jaguar and eagle imagery, hide and feathers, symbolized royalty and were used to decorate the thrones, sometimes called "eagle seat", of powerful rulers.[43] The speed, agility, and puissance of the ferocious golden eagle illustrated the noble characteristics with which the Aztec noblesse, from warriors to kings, could identify themselves.[41] Eagles and jaguars were also distinct signs of war, and elite warriors decorated their outfits with eagle feathers or jaguar skins to highlight their fearsome fighter qualities. The eagle as a war symbol went so far that the term eagle became a synonym for warrior, the battlefield was referred to as "eagle-city", and war cries of fighting soldiers were called "eagle scream".[44] Aztecs also used the eagle symbol in religious rites of human sacrifice, where the heart of the victim was removed and placed in a so-called "eagle-vessel" as an offering to the gods.[45] The victim then turned into an "eagle man" that had its place in the sky and did not need to reside with the dead. Besides describing noble features of warriors and rulers, the eagle term was also used for women that died giving birth, to honor their incredible battle. The jaguar-eagle iconography in warfare and kingship extended well beyond the Aztecs' heartland and was also used by other Mesoamerican cultures that celebrated the animals' characteristics of speed, agility, power, and wildness.

Remaining in the recognition
of the absolute view,
you should leave this life like an eagle
soaring up into the blue sky.

DILGO KHYENTSE RINPOCHE \
1910 — 1991

The Maya, a sophisticated civilization of Mesoamerica that lasted for almost 4000 years, from around 2000 BC until the Spanish conquest of their territories in the 16th and 17th centuries, also developed an important religious cult around the eagle and the jaguar, comparable to the Aztecs.[41] They inhabited Southern Mexico (today's Mexican provinces of Chiapas, Tabasco, Campeche, and Yucatan), as well as today's Belize and Guatemala, and used the two most powerful animals as formidable symbols in warfare and religious ceremonies. Their warriors dressed in eagle costumes to attack the enemy in heroic battles, and priests wore eagle attire in religious practices of human sacrifice, where they removed the hearts of captive soldiers, or members of their community, as offerings to their gods.[41]

In the early Maya cultures of the classic and post-classic period, eagles were also believed to bring rain or flood, or to evoke good and evil forces to solve daily problems.[41, 46] To the present day, Maya people use the classic eagle motif in many weaving designs or other artistic handicraft and have also incorporated the motif of a double-headed eagle in their designs. Whether this is a mythological representation of the eagle from their old cultural beliefs, or whether the image was adopted from the Spanish during the colonial time, is still under debate.[41] An oral and still current Maya tradition describes giant eagles swooping down from the sky to take humans working in the fields away to their nest or caves where they feed them to their eaglets.[41] Many versions of this myth are narrated and often include a hero of the village that outsmarts the rapacious eagles with a trick and manages to kill them. Sometimes the field workers use wooden cages or baskets that they lightly strap onto their backs so that the attacking eagle would just grab these decoys and leave the human being behind. The image of the giant eagle in this oral Maya tradition might have initially referred to Aztec eagle warriors that attacked the innocent peasants and took them back to their temples as human sacrifices. In summary, powerful eagles like the harpy eagle or the golden eagle were greatly respected in Mesoamerican cultures, and therefore omnipresent in their daily political and religious practices, reflecting warfare, noblesse, or divine power.

Eagles in heraldry around the world

Emperors, kings, rulers, and nations have always highlighted their demand for power with representations of the king of all birds. Accordingly, eagles have been depicted in coats of arms throughout history to represent and illustrate a leader's vision, courage and power. After the lion, the eagle is the second most used heraldic animal.[47] It is currently depicted on the national flag of at least seven countries (Albania, Mexico, Serbia, Moldova, Egypt, Kazakhstan, and Montenegro) as well as on the flags of the United States' unincorporated territories, American Samoa and United States Virgin Islands.[48] Technically, the eagle is found in the heraldic elements of the charge in the escutcheon, as supporter, the symbolic elements left and right of the shield, and as crest on top of the helmet.[49]

High over the oaks soar the eagle,
high over the eagle sweep the clouds.

HEINRICH HEINE \
1797 — 1856

Heraldic eagles exist in many variations, and their depictions, or attitude, evolved over time. Eagles are shown either sidewise or from the front and with wings closed, half-open, or fully displayed. The bird's head can appear single-, double-, or even triple-headed and be turned to the left or right side, or facing the observer. The animals are presented seated, while taking off, or in full flight, with or without beak and talons, and even decapitated, or as juvenile birds. Even the depiction of wing edges, tail feathers, and the angle of leg positions have different meanings and can therefore appear in many variations. Besides the complex symbolism of the eagle's varied appearance, its coloration also signals different meanings. Eagles were often depicted in black with their weapons, beak and talons, in gold and with a red tongue.

The use of an eagle in the coat of arms of European kingdoms and nations goes back to the Roman Empire, where the eagle was introduced as emblem of the Roman legions, and to John the Evangelist whose symbol was an eagle.[21, 29] Interestingly, the eagle then seems to have disappeared as a Roman imperial symbol in Western Europe for several centuries until Carolingian Emperor Charlemagne started to use it again in the early 9th century. The first coins from Western Europe depicting an eagle were minted from 1152 to 1190 during the reign of Emperor Friedrich Barbarossa. It was the Holy Roman Emperors of the Ottonian dynasty that reintroduced the eagle as their imperial symbol in the High Middle Ages.[26] At first, the Imperial Eagle was single-headed, emperor Otto III had an eagle on his scepter; but during the regency of Otto IV (1198–1218 AD) the famous double-headed Reichsadler made its first appearance on the coat of arms of the Holy Roman Empire.[50] From then on, the banner with either a single- or double-headed eagle turned into a powerful symbol of this large empire, stretching at its greatest extent from the Baltic, over Poland and Germany, to Central Italy, and lasting from the Early Middle Ages to Napoleonic times. From the mid-15th century onwards, the Habsburg dynasty consistently used the double-headed Reichsadler to represent the Emperor and the Holy Roman Empire.[51] Notably, the eagle was also used by the nobility as part of family emblems without referring to the Roman Empire or Christian symbolism. This was a common practice that started in the 12th century and lasted throughout medieval times.

Although the double-headed eagle was already present much earlier in Byzantium, the mighty eagle symbol appeared from the 15th century onwards as official imperial emblem of the Eastern Roman emperors belonging to the Palaiologos dynasty, around the same time that it was introduced by the Holy Roman Empire in Western Europe.[52] Referred to also as the Byzantine Empire, the former eastern provinces of the Roman Empire became the most powerful entity in Europe after the disintegration of the Western Roman Empire in the 5th century. The Byzantine Empire, with its capital Constantinople, lasted for almost one thousand years until its fall in the 15th century when the Ottomans took over. In the context of the Byzantine Empire, the double-headed eagle symbolized the two Roman Empires with the head on the left representing Rome in the West and the head on the right, Constantinople in the East. Also, the Balkan rulers of Serbia, Bulgaria, and Albania have used the double-headed eagle since the 13th century. Further north, the Grand Principality of Moscow incorporated the Byzantine emblem into their coat of arms when Ivan III married Zoe Sophia Palaiologina, the official heir to the Byzantine Palaiologos dynasty, in the year 1472.[53, 54] Already from the 13th century, Czech and Polish rulers also started using the eagle in their coats of arms, and in both countries a single-headed eagle is still present in the modern version of their national symbols.[55, 56]

Based on the example of the *Aquila* standard of the Roman legions, Napoleon introduced in 1804 a similar figure of a gold-cast or gilded eagle, mounted on a staff, as the military standard of his Grand Armée.[57] Referred to as the French Imperial Eagle *(aigle de drapeau)*, the eagle standard was carried into battle as regimental symbol during all Napoleonic Wars, with the aim of inspiring pride, courage, and loyalty in his soldiers. Similarly to Roman times, the standards were defended at all costs, and the loss of one of these important war symbols to an enemy was a great dishonor. Nevertheless, in several battles of the Napoleonic Wars the Russians, the British, and the Spanish defeated Napoleon's army and captured French eagle standards.[58]

> Really great minds build their nests, like eagles, in lofty solitude.
>
> ARTHUR SCHOPENHAUER \
> 1788 — 1860

Given the very long history of eagles in heraldry, dating back to Roman times, it is no surprise that the mighty raptor is still used in the national coat of arms of many countries today.[59] Russia, Serbia, Albania, and Montenegro for example have variations of a double-headed eagle as their national symbol while Germany, Austria, Czech Republic, Moldova, Poland, Romania, Armenia, and Liechtenstein feature a single-headed version. Iceland uses an eagle as supporter in their coat of arms, while the Mexican version features the famous snake-eating golden eagle that originates in Aztec traditions. Panama on the other hand uses a rising harpy eagle in place of a crest in their coat of arms, while the African countries of Ghana and South Sudan have eagles as supporters in their national emblem. Also, Nigeria and Namibia feature eagles in their coats of arms but these countries depict raptors in place of a crest. In a blend of a naturalistic and heraldic depiction of a bald eagle, the United States of America had already by 1782 incorporated their endemic mighty raptor in their Great Seal.[60] Omnipresent on flags and seals of the president, military organizations, postal services and corporate logos, the popular emblem has become a synonym for the proud American nation and their values.[61] Interestingly, the bald eagle of the United States has made its way also into the coat of arms of the Philippines, where it decorates the shield together with the lion from Spain, in memory of the Spanish and United States' colonial past. Also, in the Arabian world, several nations chose an eagle as part of their coat of arms. Originating in the Eagle of Saladin, a depiction of an Egyptian vulture appeared on the flags during Saladin's reign over Egypt in the late 12th century, the eagle was again introduced as national emblem of Egypt in the mid-20th century.[59, 62] Currently, the single-headed Arab eagle is used in the coat of arms of Egypt, Iraq, and Palestine.[59]

Besides the use of eagles as national emblems, many nations chose the powerful raptor as their official national bird. *Aquila chrysaetos*, the golden eagle, became the national bird of Mexico, Scotland, Serbia, and Albania. Indonesia chose the Javan hawk eagle as their national bird, while Namibia, Zambia, and Zimbabwe selected the African fish eagle. Panama on the other hand assigned the harpy eagle, the Philippines the endemic Philippine eagle, Poland the white-tailed eagle, and the Unites States chose the bald eagle as their national bird.[63, 64]

Given that golden eagles were once present throughout Switzerland, it is not surprising that the king of the sky found its way into many flags and coats of arms of Swiss cantons, cities, and villages.[65] Despite the ubiquity of eagles in the Middle Ages, a much greater influence that contributed to the choice of an eagle in the coat of arms must have been political considerations or the fact that Swiss territories were part of the Holy Roman Empire. The Canton of Geneva has a black half-eagle with a red crown, beak, tongue, and talons in its flag that goes back to the days when Geneva was part of the Holy Roman Empire. Similarly, a black half-eagle is depicted in the coat of arms of Altdorf, the capital of Canton Uri in Central Switzerland, probably expressing in a similar way its membership of the Holy Roman Empire of the German Nation. Aarau, the capital of Canton Aargau, has a complete black eagle in its flag which probably originates in a local oral tradition referring to the place of the eagle. That the coat of arms of the Swiss villages of Aquila (meaning eagle in Italian), located in the Canton of Ticino, and Aigle (meaning eagle in French), located in the Canton of Vaud, depict eagles is not too surprising. However, the fact that altogether almost 100 Swiss cities and villages use eagles in one way or another in their flags is rather astonishing.[66] From double-headed eagles (e.g. Simplon Dorf, Canton of Valais), through Griffins (e.g. Choulex, Canton of Jura), to half-eagles (e.g. La Verrerie, Canton of Jura or Kirchberg, Canton of St. Gallen), or just an eagle head (e.g. Filisur, Canton of Grisons), the diversity of eagle depictions in coats of arms of Swiss cities and villages is enormous.

THE LIFE OF GOLDEN EAGLES

Introduction

The majestic golden eagle has its place among the most famous raptors of this planet. The king of all birds has earned its glorious reputation due to its wide geographical distribution, impressive size, noble characteristics, and cultural and mythological significance. It is the most widespread eagle of the northern hemisphere and most likely the best studied of all eagle species.

A. chrysaetos is found over the entire Eastern European and Asian landmass, which still provide vast untouched wildernesses where the large predator birds can flourish. From the Baltic to Turkey in the West over Siberia and onwards to China, Korea, Japan, and the Kamchatka peninsula in the East, golden eagles populate most parts of the Eurasian continent. Exceptions are the Indian subcontinent (south of the Himalayas), Southeast Asia, and the Tibetan Plateau north of the Himalayas to the Gobi Desert, where the large bird of prey is absent. Only small populations are present in North Africa (Morocco, Algeria, Tunisia, Libya), the Sahara (Mauritania, Mali, Niger), in Ethiopia (Balé Mountains), as well as on the Arabian Peninsula. In contrast, golden eagles are still quite common in the vast wilderness of the North American continent from Alaska to Mexico. In Europe, golden eagles are found in parts of Portugal, in most of Spain, in the mountains of Italy (Apennine, Alps) including the islands of Sardinia and Sicily, over the Balkan peninsula, in the mountainous regions of Southern France (including Corsica), as well as in the Alps of Switzerland, Germany, and Austria. Populations in Northern Europe, on the other hand, inhabit parts of Scotland, and are found all over Norway, Sweden, and in Northern Finland.[67]

Golden eagles populate very diverse habitats, where open and half-open territories provide sufficient prey, and nesting opportunities in cliffs or trees are plentiful.[68] Habitats vary from alpine mountain ranges in Central Europe (or mountain ranges in the Caucasus and the Himalayas in Asia) to marshland with forests in Scandinavia, and to semi-arid landscapes in Northern Africa. Although the raptors normally prefer mountain ranges, they also populate lowlands if a suitable habitat satisfies their needs. A common requirement for all habitats is their demand for virtually untouched wilderness. Therefore, golden eagles do not inhabit developed areas and avoid regions with intensive agriculture. Well-adapted to a cold climate and harsh conditions, golden eagles are normally sedentary and occupy their territory all year round. Exceptions are populations that live at latitudes 50° north or greater, which are forced to migrate south because of decreasing availability of food in winter. Normally these populations just migrate a few thousand kilometers south where they spend the roughest months of the winter. Such migratory behavior of golden eagles has been observed in the far north of Finland, Russia, Alaska in the USA, and Canada.[68, 69, 70]

The species of the golden eagle has been well established using traditional ornithological methods such as description of size, body features, coloration, or behavior. The Swedish botanist, zoologist, and physician Carl Linnaeus (1707–1778, later Carl von Linné) formally introduced the binomial nomenclature of modern taxonomy for living organisms, and first described the golden eagle in the tenth edition (published in 1758) of his monumental work *Systema Naturae* as *Falco chrysaetos.*[71] Two years later, moved to the genus *Aquila*, the golden eagle received its current systematic name *Aquila chrysaetos* from the French ornithologist M.J. Brisson in the year 1760.[72] While *Aquila* is the Latin word for eagle that likely originates in *aquilus*, meaning "dark colored", *chrysaetos* has its roots in *khrūsós* meaning "gold" and *aetos* meaning "eagle" in

Ancient Greek.[73] The systematics of the golden eagle has been complemented with the assignment of the *Aquila* genus to the *Aquilinae* subfamily of the *Accipitridae*, a family that classifies more than 230 bird species. Molecular genetic investigations found the golden eagle to be most closely related to the following: Verreaux's eagle (*Aquila verreauxii*), a large black eagle present in Southern Africa; Gurney's eagle (*Aquila gurneyi*), a large eagle mainly found in New Guinea; the wedge-tailed eagle (*Aquila audax*) from Australia; the widely distributed Bonelli's eagle (*Aquila fasciata*); and the African hawk-eagle (*Aquila spilogaster*).[2]

Based on body size, coloration of plumage and molecular phylogenetic analyses, golden eagles are further divided into six subspecies, although it is debated whether some of them could be grouped together.[68, 74]

— *A. c. chrysaetos*: The nominate species introduced by Linnaeus in 1758 which is also referred to as the European golden eagle. It is a medium-sized golden eagle with the lightest brown plumage of all subspecies and featuring a splendid golden nape. Individuals are found in Scotland and Northern England, in Scandinavia, and in the Alps of France, Italy, Switzerland, Germany, and Austria. In Eastern Europe, it populates territories in the Baltic, and over Western Ukraine, Romania, and Bulgaria, to Greece and the Balkan countries. It is also found in Western to Central Russia, Kazakhstan, Uzbekistan, Turkmenistan, and in Northern Iran.[68, 75]
— *A. c. daphanea*: Also named Himalayan golden eagle or berkut in Russian, this subspecies describes the largest golden eagle that is mainly found in Central Asia, over the Himalayas, to West and Central China. With wingspans averaging over 2.2 m, these large eagles have a darker blackish plumage on the back and a brownish-red nape. They soar over the mountains of the Caucasus and the Himalayas and are also found in Mongolia, Central and Eastern China.[68, 74]
— *A. c. homeyeri:* A slightly smaller and darker subspecies than *A. c. chrysaetos* that is found in Spain, Portugal, and the Greek island of Crete. Further east, territories extend through Turkey, the Middle East, the Arabian Peninsula, mountains of the Caucasus, and from Iran to Southwestern Kazakhstan. They are also present in North Africa near the Mediterranean coast of Morocco, Algeria, and Tunisia, while an isolated population is present in Southeast Ethiopia in the massif of the Balé Mountains.[68, 75]
— *A. c. japonica*: Commonly named Japanese golden eagle, it is the smallest and darkest subspecies of *A. chrysaetos* that soars in the mountains of Honshu and Hokkaido in Japan and in parts of Korea.[68, 75]
— *A. c. kamtschatica*: A large subspecies that populates most of Russian's Siberia, including the Kamchatka Peninsula in the east that gave the subspecies its name. It is also found in Mongolia. Although bigger, individuals of *A. c. kamtschatica* are very similar to golden eagles found in North America. It is therefore debated whether they really form an independent subspecies.[68]
— *A. c. canadensis*: A medium-sized, dark subspecies of the North American continent from Alaska to Central Mexico that is also darker and smaller than *A. c. chrysaetos*. This subspecies is well known as the American golden eagle that plays a highly important role in the cultures of the First Nations people in Canada, the Native Americans in the United States, and the indigenous cultures of Mexico.[32, 33, 41, 68]

Anatomy and physiology

Adult golden eagles have a dark brown plumage, with patches of gray on the inner wing feathers, and a typical golden collar around the neck, giving rise to their common species name. Both male and female possess similar coloration of their plumage. The skin around nostrils and mouth, the cere, the orbital ring around the eyes, as well as the skin on the three forward and one rear-facing digits, are bright yellow. The large beak, with downwards-pointing hook, is of gray color at the base and gradually changes to deep black at the tip. The formidable weapon of the golden eagle, with which it catches and kills prey, are the strong and sharp curved talons, up to 6.3 cm long, and also colored in deep black.[76] The eagle's entire, long legs including the tarsometatarsus, the fused ankle/foot bones that are typical for birds, are completely covered in feathers and therefore seldom visible. The broad tail of the raptor is also dark brown but often contains diffuse, discontinuous bands of lighter color.[74]

Golden eagles are very large birds of prey with body sizes up to 1 m and large, broad wings with 11 primary and 17 secondary flight feathers while the tail has 12 steering feathers. The primaries, up to 60 cm long, fan out and bend upwards, giving the raptor, together with the slight V-shape of the wings, a highly typical appearance in flight.[75, 77]

Male and female birds differ significantly in size and weight.[76] Similarly to other eagle species, female birds are bigger, reaching a body size of around 1 m and a wingspan up to 2.3 m.[74] A male's body is around ten centimeters shorter and the wingspan reaches up to 2.1 m. Golden eagles are heavy birds, with females weighing up to 6.7 kg, while the smaller males reach a maximum weight of about 4.6 kg.[74] Different sizes between male and female, with significantly larger females, is a very common feature in raptor species. It is referred to as inverted sexual dimorphism and thought to be most prominent in species that hunt highly agile prey. Especially in the first period of the breeding cycle, the smaller size and weight of the male guarantees increased agility in difficult hunting terrain and higher mobility in using thermals, which together result in greater probability of killing prey to feed the brood. The female with its much bigger size and weight, on the other hand, is much more suited for the long incubation period of the eggs and the early days of upbringing the young in the nest, where it needs to be immobile for long periods and profit from larger fat depots in the bigger body.[68] Under favorable conditions, golden eagles can live very long, with documented maximum ages up to 32 years.[68] Despite their very impressive appearance, golden eagles have an unspectacular, short call resembling a shrill *seeir* that they might repeat several times. Although well known for their silence, the mighty raptors have developed several distinct, additional vocalizations that they use for male to female communication or when they interact with their young. These varying vocalizations are specific to different contexts such as nest-building, food deliveries, alarm calls, localization of the partner, or arrival at the nest, but are rarely audible. Nestlings regularly express food-begging calls that develop from a chirping sound of fresh hatchlings, to a two-syllable tsik, and then to a call that closely resembles an adult *seeir* vocalization.[74]

Juvenile golden eagles are darker brown during their first year and have a very typical white coloration on the basis of the inner primaries and outer secondaries on the remiges of the wings, as well as a sharp white band at the base of the tail feathers. These easily recognizable white patches are then slowly replaced by darker pigmentation over several years and the final adult appearance is reached at an age of five to seven years.[68, 74, 78] Similarly, the cere changes from a very bright yellow in juvenile animals to a much duller white-yellow color in adult animals. Golden eagles regularly renew their plumage during molt in the summer months. The process starts at the head and neck, and slowly progresses over the entire body to the tail. Wing feathers are replaced from inner to outer feathers.[68, 79]

It is not necessary for eagles to be crows.

SITTING BULL
(TȞATȞÁŊKA ÍYOTAKE IN LAKOTA) \
1831—1890

Golden eagles are often referred to as the kings of the sky due to their exceptional flight skills, that are among the most developed of all birds of prey.[80] Large, broad wings together with a wide tail amount to a substantial surface area that makes them excellent gliders despite the rather heavy body weight. Most of the time, golden eagles effortlessly soar in thermal updrafts with fully outstretched wings kept in slight V-form, and hand feathers spread and bent upwards.[68, 77] The tail is aligned with the wings or turned sideways and up or down for steering. In fast maneuvers, especially for braking, eagles can spread their tail feathers to almost double the surface. If there is not enough uplift, eagles use flapping flight modes where they alternate half a dozen wing-beats with a few seconds of gliding flight.[68] Since flapping flight costs a lot of energy for the heavy raptors, they keep flapping to a minimum and rather wait on a perch for the sun to warm up the air and produce thermals. For faster flights than just gliding, golden eagles reduce the wing surface gradually from slightly folded wings for intermediate speed to tightly closed wings during nosedives at which the predators can reach enormous speeds, far above 250 km/h. In all normal flight modes, the raptors usually keep their legs and feet aligned with the body along the tail. Only during takeoff and landing, before and while catching prey, in a parachuting vertical downwards flight, or in territorial fights with other eagles or different intruders, do they stretch out their long legs, with fearsome talons, in a position more or less perpendicular to the body and plane of the wings.[74]

Golden eagles and other diurnal raptors from the *Accipitriformes* or *Falconiformes* orders are well known for their phenomenal eyesight, detecting prey with high precision from several kilometers away. For these birds, vision is the most developed sense, and essential for hunting and navigation through dense obstacles in fast flight. The importance of vision for birds is reflected in the size of the ocular globes (eyeballs) which are often larger than the bird's brain. An eagle's eye, being up to 3 cm in axial length, is approximately the same absolute size as a human eye, although the raptor's body mass is 10–20 times smaller, a finding that again emphasizes the importance of vision in raptors. The golden eagle's iris is of brown to hazel (or brown-gold speckled) color with a round black pupil, while the sklera is white as in a human eye; but in the eagle's case, covered by the eyelids, it is not visible.[74] Besides the upper and lower eyelid, golden eagles have, as many other birds, a third translucent eyelid called the nictitating membrane that can be drawn across the eye from the corner near the beak for protection, cleaning, and moistening while maintaining vision. In contrast to humans, eagles blink upwards since their lower eyelid is larger than the upper one.[81]

Eagle eyes are high-tech evolutionary developments endowed with some of the highest visual acuity of the animal kingdom that outperform the capabilities of human eyes by far.[82] The extraordinary eyesight of raptors is achieved by several specialized anatomical modifications in the eye that can be found in all three components that define the optical performance of the camera-type vision of vertebrates: the optical system (lens), image analysis (retina), and the visual field (position of eyes). The long axial length of the eagle eye results in a long focal length that projects a large image onto the retina, contributing together with the wide pupil to an acute vision and high spatial resolution.[83] In contrast to humans, who can only actively change the curvature of the eye lens for focusing, several birds of prey can also adjust the curvature of the cornea to increase the refractive power. This second accommodation system allows golden eagles and related species to focus independently on two different objects with each eye, using monocular vision in the separate, non-overlapping field of view.[83]

As in all vertebrates the eagle retina is inverted, meaning that the incoming light first passes through several layers of neuronal cells and optic nerve fibers before it hits the photoreceptors.[84] Diurnal raptors usually have the complete set of photoreceptors found in birds: four different cone receptors and a double cone receptor for color vision as well as rod-type receptors that are sensitive to dim light. The cone types differ in the spectral sensitivity of the photon-absorbing retinal pigment (opsins) and the composition of small oil droplets that act as optical filters.[83] The detail with which an optical system can resolve an image is defined as spatial resolution, which is dependent on the photoreceptor density. A high density results in a higher resolving power for small details in an image. Therefore, raptors have rather thin photoreceptors that can be packed very densely, and both the thicker double cones and dim-light sensitive rods are absent in some raptor species in high acuity regions of the retina for denser packing. In contrast to humans, which have one invaginated circular spot where photoreceptor density is increased, the fovea, eagles possess two such regions which are even connected by a narrow, elongated area also containing increased photoreceptor cell numbers. These eagle eye foveae, referred to as central fovea (deep invagination) and temporal fovea (shallow invagination), reach much higher receptor densities compared to a human fovea, and are partially responsible for the much sharper eagle eye vision at a similar, narrow visual field.[83] While a human fovea has around 150,000 light receptor cells per square millimeter, the receptor density of an eagle fovea is several times higher, resulting in around two times higher resolution.[85, 86] The precise function of the two raptor foveae is still not well understood, and especially the optical performance of the deep fovea with its distinct shape is debated among scientists. It has been suggested that it serves either as a local image enlarger or as a focus indicator and movement detector.[83, 87]

Another anatomical adaptation to the bird's eye is their more side-wise position on the skull that results in a very different visual field compared with humans. Humans have both eyes in front of the skull yielding a large overlap of each eye's visual field (~120°) for an optimal stereoscopic binocular vision. In birds, each eye looks outwards at a different scene, resulting in a very wide total visual field that can sometimes cover more than 300° and the overlap of each eye's visual field for binocular vision is small (~20–47° depending on species).[83] Binocular (stereoscopic) vision towards the front of the raptor's head is thought to be needed for optimal control and timing of bill and talons for catching prey, while the monocular side vision is optimized for the detection of prey or predators.[83] This interpretation of a raptor's visual system makes sense if one considers that they usually keep their head straight longer to look at close objects using their forward-pointing temporal fovea in stereoscopic binocular vision, while they almost exclusively use sideway monocular vision for distant objects that they focus with the more outwards directed and much denser central fovea to achieve most acute sight.[83, 88] Given the large visual field, it is understandable that raptors have a much smaller degree of eye movement compared with humans, although they have a surprisingly flexible neck that can turn the head over a much larger range than humans.[83]

Since the visual system of golden eagles is optimized for daylight activities, their visual performance quickly drops as light levels decrease. Interestingly, it has been shown that the contrast sensitivity of raptor vision is lower compared with many mammals including humans, which are themselves exceptionally good at this. While it is clear that eagles with their four different cone types must have excellent color vision, their eyes lack significant transmittance of short wavelength radiation meaning that they probably have very limited sensitivity to UV light.[83]

Golden eagles and related species have a very impressive fierce gaze that has fascinated humans throughout history. Besides the piercing eyes, it is the raptor-typical bone ridge above the eyes that contributes to this wild and noble look. These protruding bone structures are thought to serve as sun shades to prevent the sun being directly imaged on the retina, which would lead to glare and a worse optical performance.[83]

Golden eagles in Switzerland

The golden eagle is one of the five large predators that had been naturally present in Switzerland over many centuries. While the brown bear (*Ursus arctos*), the wolf (*Canis lupus*), the lynx (*Lynx lynx*) and the bearded vulture (*Gypaetus barbatus*) had been hunted to extinction by relentless persecution, small populations of the golden eagle could survive in remote corners of the Alps. Hunting and looting of nests led golden eagle populations in Switzerland to drop to a minimum at the beginning of the 20th century, a time when dozens of golden eagles were killed every year.

Hunters were encouraged by official government laws to shoot as many individuals as possible and received premiums for the killing of what was believed to be livestock damaging animals such as the golden eagle. The remaining small populations did not recover even after 1925, when Switzerland issued new hunting restrictions that prohibited the killing of an eagle near the eyrie or looting of eggs and nestlings. Only after a nationwide, complete ban of persecution in the year 1953 did population numbers slowly grow again. In the last 70 years, the golden eagle population steadily increased in Switzerland and has stabilized at around 350–360 breeding pairs, allowing occupation at each appropriate territory.

Currently, population numbers are stable and are regulated by intra-species competition for the finite number of territories.[89, 90] Golden eagles occupy territories in the entire region of the Swiss Alps, over all foothills of the Alps, and the first pairs have been sighted breeding in the hills of the Jura in Northwestern Switzerland. This means that golden eagles have established home ranges all over Switzerland except for the large arch of the so-called Swiss Midlands that extends from Geneva, Berne and Zurich to St. Gallen and that features extremely dense infrastructure, high population numbers, and intensive agriculture, characteristics that are definitely not compatible with the needs of golden eagles. The re-establishment of a healthy and stable golden eagle population in the mountains of Switzerland is a great success for species conservation. It also provides an excellent opportunity to study the population development and regulation of a large predator species in the Alps and gives valuable insight for the reintroduction and recovery of other highly endangered species such as the bear or the wolf, both of which once naturally roamed our mountains and valleys. In addition, iconic predators such as the golden eagle, bear and wolf, and also the lynx, have a strong symbolic significance for the biological integrity of the Alps and are therefore attractive species to showcase its healthy natural state. Such examples will positively contribute to the reputation and attraction of Switzerland, a small and mountainous country in the Alps, for which tourism is a highly important income factor. If the golden eagle helps to attract nature lovers interested in eco-friendly, close-to-nature tourism experiences, not only the people of Switzerland but also all the wildlife, and nature in general, will benefit.

Territorial behavior of golden eagles

A golden eagle breeding pair occupies a defined geographical territory for many years and defends it relentlessly from intruding single sub-adult individuals that are passing by. A study in the Grisons, an eastern part of Switzerland full of rugged mountains in the massif of the Alps, found that territory size varies from 28 to 90 km^2, while studies in different areas of the Alps uncovered territories as large as 200 km^2.[89, 91] Accordingly, population density in the Alps varies somewhere between 5–45 breeding pairs per 1000 km^2 depending on topographical characteristics, availability of nesting sites, and abundance of prey. Golden eagles of the Balé Mountains in Ethiopia and of Idaho in the United States on the other hand seem to get along with much smaller territories of roughly 1–10 km^2 and 12–50 km^2, respectively, likely reflecting a very high density of prey.[92, 93]

Eagles don't catch flies.

DESIDERIUS ERASMUS VON ROTTERDAM \
CA. 1466 — 1536

Ideal golden eagle territories feature more than 50 % open or half-open landscapes that are sparsely wooded to enable hunting of prey. Given the wide global distribution of *A. chrysaetos*, habitats greatly differ from tundra, over shrubs, scrubs, and grasslands, to coniferous forests, and high alpine zones.[74] Often, these habitats lie in inaccessible regions where agriculture and commercial forestry are kept to a minimum.[89] In mountainous regions, territories are bordered by mountain ridges and therefore form naturally defined spatial areas usually consisting of a few connected valleys.

In the Alps, home ranges of the golden eagle span several vegetation zones, from the low altitude arborous mountain zone, over the subalpine zone, to the alpine zone. Although golden eagles might visit the low-level mountain zone with its deciduous forests, they spend most of their time in the upper subalpine zone near the tree line, where open conifer forests and isolated trees provide ideal hunting and nesting grounds. The alpine zone with its vast meadows and gravel fields is another preferred hunting area since prey can be detected easily and transported downwards to the eyrie, which is often built in the subalpine zone. Under ideal conditions such as strong winds or extreme thermals in the hot summer months, golden eagles also visit the nival zone with its bare rocks and eternal snow. They can reach surprisingly high altitudes of up to 4500 m above sea level, where they have been sighted by glider plane pilots or mountaineers.[89] In general, golden eagles spend most of their time on the slopes of the valleys and peaks between ~1500 to 3000 m above sea level, trying to detect prey or to patrol their territory.

Besides the availability of prey, the ideal golden eagle territory needs to provide several suitable nesting sites. In the Alps, golden eagles most often build their nests on ledges or in small caves in inaccessible cliffs, and only sometimes use old pine trees. Golden eagles construct eyries using tree branches that are piled up several meters in height. As the same nest is often used over many years by subsequent generations, the nest can become several hundred years old before it finally collapses or is given up in favor of another nesting site.[89] A golden eagle pair usually maintains several nests in its home range that are used alternately. Geographical orientation of nesting sites is often downwind, and eyries are located several hundred meters above the valley floor to enable easy take-off and landing. Importantly, nests lie below main hunting grounds since the transport of prey becomes much easier downwards than upwards. The large predators seem to have no preference for the cardinal direction of the nesting site, but like to choose sites that are well protected from the weather, often in caves or on ledges under cliff overhangs.[74, 94]

Depending on the season, weather, and breeding activities, the behavior of golden eagles changes significantly. Studies in Switzerland and the Unites States found that nesting resident golden eagles raising offspring spend around 75 % of their time on the nest or regularly used perches.[89, 93] Flight time follows a clear daily rhythm relating to the position of the sun which creates local up-winds that are used by the predator birds for elevation gain. In the morning hours, eagles fly along east or southeast facing slopes or cliffs where the first thermals develop, and rarely visit parts of the valleys which are in shade. This changes over mid-day and early afternoon when they visit almost all regions of their territory, as updrafts develop in any orientation and are peaking. Later in the afternoon and in evening, they spend most time on southwest or west-facing valley slopes to make use of thermals generated by the evening sun.[89] Flight activities are greatly dependent on the weather. On sunny days, eagles follow the daily flight pattern to visit each area of their territory, but on rainy days when hardly any up-winds develop, they spend most of their time sitting on perches. Winter and summer activities also differ significantly. In winter time when thermal updrafts are weak due to the low position of the sun, golden eagles are only active for a couple of hours around noon and preferably visit sun-illuminated south-facing cliffs and mountain slopes; while during the summer months, with their strong thermals, eagles can be seen in the air at any time of day. Besides thermal upwinds, golden eagles use wind systems to get uplift. If a constant air current hits a mountain slope or cliff, the air is deflected and forced upwards where it produces a wave-like turbulence on top of the mountain or cliff edge that can reach high up into the air and extend far above peaks or ridges. Golden eagles extensively use such wind phenomena, referred to as orographic updrafts, to reach higher altitude, soar along ridges, or even remain still at one point in the sky to scan the surroundings for prey.

> Fierce eagles do not breed the tender dove.
>
> QUINTUS HORATIUS FLACCHUS \
> 65 — 8 BC

Flight activities also differ during low pressure and high pressure weather periods. Low pressure systems or fronts generally produce strong winds that are used to carry the birds at any season high up into the air, whereas stable, long-lasting high pressure periods with their weaker daily wind systems reduce flight activities, especially in winter.

Resident eagle pairs defend their home range aggressively and do not allow other pairs or single eagles to enter their territory. Due to their outstanding eyesight, invading individuals are spotted by the territorial animals from very far and are immediately attacked. The residents reach the invaders in long gliding flights, supported by wing flapping if needed, or directly attack in spectacular nosedives if they fly above the invader. Most of the time there is no direct physical contact and the invader will be chased out of the home range, either by intimidation via undulating territorial flights or direct chase behavior. If eagles are getting really close to a fight, the bird being attacked will roll onto its back and present its talons ready to defend itself if needed. Rarely, serious fights develop that can result in an entanglement of both eagles, when the attacker also uses its talons to get hold of the invading individual. Entangled with each other, both eagles can then coast freely down for several seconds before one of them finally releases its grip just above ground. Single patrolling eagles of the resident pair carry out these attacks on their own but, if in company with their mate, it can be that both of them strike to chase the invader out.[74, 89]

Defending their territory against invading single eagles is stressful for the resident pair as it costs a lot of energy and time. In extreme cases, when conflicts with other individuals become too regular, it negatively influences breeding success and is therefore a factor that naturally regulates eagle numbers and defines the appropriate size of a territory.[89]

In the Alps, young eagles usually start migration during their first winter, between November and April. In the following two years these sub-adult eagles roam vast areas as, unlike breeding pairs, they are not bound to nesting sites. Visited areas can be as large as 15,000 km^2 for young males and slightly smaller for young females.[89] This often leads to encounters with other ranging eagles, but, in contrast to territorial breeding pairs, single eagles tolerate each other. Older single eagles (3–5 years of age) then continuously reduce visited areas to ~3500 km^2, usually around the place where they were born, until they find a mate with which they finally become territorial.[89]

Hunting strategies and feeding

Golden eagles are excellent predators that feed on a wide variety of prey, depending on availability. From small birds and rodents, to hare, marmot, and foxes, up to much larger animals like chamois or ibex, no species escapes the sharp eyes of the golden eagle in the Swiss Alps. Verification of prey during the breeding season in the summer months has been carried out by collecting and analyzing remnants of killed species found in nests.[89] These studies from the Grisons, in the eastern part of Switzerland, discovered that marmots were the preferred prey, representing a biomass of more than 75 %, followed by chamois and doe with biomass percentages of 9 % and 4 %, respectively. The distribution changed in the winter months when marmots were hibernating. Ungulates then dominate as prey, with chamois being by far the most preferred species, followed by doe, deer, and ibex. Most often it is the young animals up to 15 kg that are taken down by golden eagles, but they also make use of adult carrion that perished due to old age or harsh winter, or individuals killed in avalanches.

Due to their incredibly powerful vision, golden eagles spot prey from several kilometers away, either during search flights or from a perch on trees or cliffs from where they scan the surroundings. As diurnal raptors, golden eagles hunt from dawn until dusk, when the predator's eyesight far exceeds the optical performance of their prey. Golden eagles have developed several hunting strategies that vary depending on the size and mobility of the animal they attack, the topography and vegetation of the landscape, and the position of the quarry relative to the eagle.[68] Despite their famed reputation as highly skilled hunters, on average only every fifth golden eagle attack was observed to be successful.[95] This number should be taken with caution as success rates might vary greatly from habitat to habitat with different availability, variety, and abundance of prey. The large raptors are able to vary the many hunting techniques back and forth during a chase, and certain techniques show individual variations from eagle to eagle.[68] Although golden eagles hunt most of the time alone, cooperative hunting of breeding pairs is a common phenomenon, but seems to be less successful.[95]

Once prey is spotted at a distance during a search flight, or from a perch, the eagle starts a low angle gliding-flight, during which it increases speed by folding its wings closer to its body. Around 20 m before reaching the prey, the predator flies with half-open wings and fans its tail to slow down, whilst stretching out the long legs with the sharp talons. The eagle still maneuvers and corrects the flight trajectory in this position to follow the animal when it changes location or flees. At this stage of the attack, eagles often fly close to the ground and use topographical elements like rocks, small hills, or trees to maximize surprise before the strike. The eagle then fully opens its wings and fans its tail right before impact, grabbing the victim before piercing it to death with its talons. This method is particularly useful for marmots, hare, grouse, or even ungulates that roam in semi-open landscape and that are spotted from far. In another method, eagles surprise their prey by low level contour flights close to the ground, so that the victim cannot see the attacker above the horizon. Using natural cover, the predator bird takes the prey by complete surprise and snatches it before the danger is realized. Eagles seem to use this method preferentially for colonial, burrowing mammals such as marmots, which have a very short escape path down into a burrow. If an attacked quarry manages to escape in fast flight, but cannot find safe cover, golden eagles also pursue their victims in a tail-chase. Due to its superior agility, the eagle often manages to match all twists and turns of the escaping animal until the raptor finally drives its talons into the victim's body. Rare cases have been observed where high-soaring golden eagles stoop with fully folded wings towards birds that fly at a lower level to catch them with high impact. In this way, golden eagles can reach speeds far above 250 km/h, so prey that initially was quite far away can be reached within seconds. This method is only successful for large and slow-flying birds like geese or grouse, as eagles are by far less agile at this very high speed when compared with a peregrine falcon that hunts fast-flying birds like pigeons or jays in a similar way. If the animal that is being attacked by the golden eagle is not very mobile, or shows a freezing reaction that keeps it still, the large bird of prey can also hover for a short time over the victim and slowly descend to finally drop down to catch it. Such a hunting strategy has been observed for snakes, hedgehogs, or even newborn ungulates that hide motionless in deep grass. Rarely, golden eagles also attack non-mobile animals by touching down close by, walking towards the victim, and snatching it or pulling it out from a hide with its talons.

In general, if an initial surprise attempt to clutch the victim fails the prey is normally lost, as a golden eagle is not very mobile on the ground. If the hunt was successful, golden eagles usually kill the quarry quickly by piercing vital body parts with their talons, but sometimes eagles can also be dragged along for several meters by powerful animals that try to escape. Often the large raptors take young mountain goats or doe by the head and kill them with the sharp talons penetrating through the skull and into the animal's brain.\89\

Impressive cases of a very bold hunting method have been observed where a golden eagle flies into a group of ungulates, choses a victim, and drives its talons deep into the back or neck of one individual. The attacked animal then panics and tries to escape in a frantic flight down steep slopes or rocky cliffs with the eagle on its back. During the wild ride the eagle stretches its wings and flaps to keep balance while it tries to cause deadly injuries with his claws. Such attacks can last several seconds or minutes until the ungulate either collapses due to injury, exhaustion, or shock, or the eagle must let go to save his own life. In certain regions, golden eagles have specialized in yet another spectacular hunting strategy, where they attack hoofed game that traverse small ledges on high cliffs or animals that stand close to a gaping abyss. The golden eagle attacks and drives its talons into the body of the victim just to carry it a few meters out into the void before it opens its claws to release the heavy body, which then drops all the way down the cliff where it perishes from the heavy impact. The eagle then quickly follows the victim to the landing position to feed or to dissect it for transporting body parts to the eyrie.\89\

Golden eagles sometimes feed for hours from killed prey, storing large amounts of food in the crop, an expandable pouch near the throat for temporary food storage. Filled crops can best be spotted on eagles in flight, where they manifest as large projecting bulbs below the throat. During nesting season, the predator birds need to hunt regularly to supply their young with food, but the large birds can easily stay without large quantities of food for several weeks when they do not have nestlings to feed. Since golden eagles cannot carry killed prey that weighs more than their body weight for long distances or upwards, they usually dismember bodies and fly with pieces to their nest or return several times to the cadaver to feed.

Golden eagles are highly devoted animals that mate for life and normally neither leave their partner nor the joint territory that they occupy together. Besides soaring with outstretched wings, short flapping flights, fast directional flights with angled wings, or straight nosedives, golden eagles also perform characteristic undulating flights. In these display flights, the raptors conduct a series of up and down movements in midair, where they stoop downwards with completely folded wings followed by slowly opening their wings which leads into an upwards flight with fully outstretched wings. Duration, frequency, and amplitude of such undulating flights vary and have been observed at many altitudes, from close to the ground to high up in the sky. These spectacular flights are performed by both male and female eagles of different ages, either when flying together or on single flights. Stunning to watch, the behavioral purpose of undulating display flights is not well understood. Interpretations from a purely nuptial or sexual function, over a joyful exercise, to a signal to keep territorial intruders away, or a compulsive behavior have all been described in the literature. The current consensus seems to be that undulating display flights of golden eagles serve as a means for bond maintenance of the eagle pair, courtship behavior before a new breeding cycle, or territorial claims against intruders of the same or different predator bird species.[68, 96] Sometimes golden eagles combine undulating courtship flights with maneuvers where they drop a stone or another item only to dive after it and catch it in midair, obviously being a flight exercise supposed to impress the partner.[68] In the Alps, territorial eagle pairs mate from late January to May with peak activities during the second half of March and into April.[89]

Golden eagle pairs maintain their eyries through almost the entire year by constantly adding branches and twigs, and padding them with grass and foliage, but increase nest building activities from February onwards into spring. A little more than 40 days after mating, the female lays one or two fertilized eggs of whitish color, with often brown spots or blotches, that weigh about 150 g and measure around 75 by 60 mm.[68] In Switzerland, eggs are usually laid at the end of March and beginning of April followed by a brooding period of 41–45 days.[74, 89] During brooding the female spends most of the time incubating the eggs while the male eagle provides food.[74]

> The eagle,
> he was lord above.
>
> WILLIAM WORDSWORTH \
> 1770 — 1850

Chirping from within the egg often announces the forthcoming of the eagle hatching. But it is frequently a long struggle and can take many hours for the chick to break the eggshell and free itself. Freshly hatched eaglets are completely covered in white down and weigh about 110 g.[97] In the first few days of a baby eagle's life, the hatchling is not very active and needs constant brooding or shading by the parents to keep the body temperature at normal levels. After around 20 days of age, when the chick has already gained more than ten times its birth weight, the eaglet is strong enough to stand up.

Between 6–15 days of age, hatched eagle chicks change the short gray-white down plumage they are born with to thicker and longer, white down feathers. This still whitish coat is then slowly replaced by dark brown plumage over the next weeks in a process whereby new and thicker feather sheaths grow from the same sockets and push the downy feathers out. It then still takes around 10 days for the brown feathers to break out from the new sheaths and become visible. This replacement process starts on the wings with the primaries and the secondaries and continues to tail feathers and scapulars. As it takes the young golden eagle several weeks to grow its complete juvenile plumage, the young raptor appears between 25–50 days of age in a blotchy white and dark pattern that greatly helps in determining the age of the nestling. The eagle finally attains the mature dark brown juvenile plumage with the typical white patches on wings and tail after 70–80 days. In a similar shift, the coloration of the talons changes from flesh-colored in the newborn to deep black in the mature juvenile eaglet, while the cere and feet change from a pale flesh-colored yellow to a very bright lemon-yellow.[68] At first, adult eagles dissect killed prey in the nest and feed their young bite by bite. But after around 50 days, the fast-growing juvenile eagles are strong enough to fragment intact prey that was deposited in the nest by their parents. Cainism, the killing of a weaker sibling by the stronger brother or sister fledgling, a compelling phenomenon in other eagle species, is not common for golden eagles and only takes place in times where nutriment is limited. In an incredible transformation that takes around 70 days, the tiny, white, and helpless golden eagle chick has turned into a tall, dark, majestic, and powerful bird of prey that alertly scans the surroundings of the nest, ready to take its first flight and start the exciting life of a proud golden eagle.

At an age of around 70 days, in Switzerland usually in the second half of July, young eagles leave the nest for the first time on their maiden flight and explore the immediate vicinity of the nest on short flights during the following two months. In the next five to eight months, juvenile raptors visit the entire territory of their parents, at first in company and supervision of parent birds and later on their own. Only after around eight to ten months have they acquired all necessary flight and hunting skills, ready to leave home grounds to roam huge areas of the Alps for the next two years.[89]

> Don't fear,
> thy youth shall be renewed
> as an eagle.
>
> AURELIUS AUGUSTINUS \
> 354 — 430

Breeding success of a resident eagle pair strongly depends on continuous breeding activities and availability of nutrition during the feeding period of the newborn eaglet.[68] The likelihood of healthy offspring therefore correlates with an intact, undisturbed habitat where flora and fauna flourish. In the Swiss mountains, abundance of prey in the breeding season also correlates with the amount of snow which is left from a long winter. The amount of snow present from March to May influences the availability of ungulate carcasses to feed on by providing more or fewer animals that perished due to harsh winters or were killed in avalanches. Strong winters with a lot of snow therefore produce plenty of food for breeding eagles but also attract invading single eagles which then trigger territorial stress responses. Frequent intrusion of foreign eagles into the territory demand interception flights by the breeding pair which leave the nest and eggs unattended. If such territorial stress occurs too often, embryos will not develop properly in the egg and the brood will be lost. Therefore, the reproduction rate in golden eagles is strongly density-dependent, which leads to plateauing of eagle numbers if every territory in a certain region is occupied. This is the case in Switzerland where golden eagle numbers have stabilized at around 350–360 pairs and breeding performance is with ~0.3–0.5 (measure of independent young individuals per breeding pair and year) relatively low.[89, 90] It means that a resident golden eagle pair in the Swiss Alps raises around 3–5 healthy independent juvenile eagles in a period of ten years.

Juvenile golden eagles mature slowly and reach sexual maturity only after around five years.[68] This has to be seen in the context of their long life expectancy, which is somewhere between 20–30 years. Age determination of golden eagles is straightforward for young individuals by inspecting the degree of white patches on tail and wing feathers in addition to the overall coloration of the top side of the wing. Once eagles possess their fully matured adult colors, age classification is much more difficult.

> The whole wide ether
> is the eagle's way.
>
> EURIPIDES \
> CA. 485 — 406 BC

A feature of high age is a more grayish sallow color of the cere which is of bright yellow coloration in younger individuals. In general, five age classes are distinguished: juvenile eagles in their first year with bright and clearly distinct white tail and wing patches (juvenile plumage); immature animals between one to three years that still possess the coloration of juvenile birds but also develop a band of lighter brown color on the top side of the wings (basic I plumage); sub-adult eagles between three to five years featuring discontinuous white wing and tail patches and pronounced top side wing bands (basic II plumage); adult 1 individuals, older than four to five years and possessing only small vestiges of white tail and wing patches besides a complete dark brown plumage with lighter top side wing bands (basic III plumage); and lastly, adult 2 individuals that have lost all features that distinguish young animals (definitive basic plumage).[74]

Conservation and outlook

While human pursuit was the main reason for reducing numbers over centuries, golden eagles now often become victims of intra-species territorial fights, leading to injuries with direct or indirect fatal consequences. Another cause of death, and still a pressing problem, is lead poisoning of golden eagles after feeding on dead carcasses that have been shot with lead-containing ammunition. This highly toxic heavy metal leads to a permanent contamination of the eagle's organs, significantly weakening the bird and initiating a slow death over months or years. Initiatives to ban lead-based ammunition have been ongoing for years in several Alpine countries but a complete ban has not yet been introduced anywhere.

In Switzerland, golden eagles are a protected species, meaning that they are not allowed to be hunted down, and are listed in the Red List of threatened breeding birds, where they are classified as vulnerable (VU).[98] Although numbers have remained stable over many years, or even slightly increased, these majestic creatures remain at risk and need special attention. While hunting or poaching are very rare, it is rather the intensive use of the Swiss mountains as a recreational and economical playground that puts pressure on golden eagle populations. Despite their incredible beauty, the Swiss mountains have always been, and still are, in the process of extensive development. For such a rugged mountain region like the Swiss Alps there is already an incredibly dense infrastructure, but further development continues year after year. Whether one likes the extensive use and continuous development of the Swiss mountains or not, everybody needs to decide for himself. But we should all acknowledge that it puts enormous pressure on flora and fauna. As a consequence, we should try very hard to protect what is left and very carefully balance long-term benefits and drawbacks of new infrastructure projects to avoid further habitat destruction or the flooding of people into remote untouched mountain areas. Mass tourism should be concentrated on very few, designated destinations while sustainable and eco-friendly individual tourism can be distributed over wide areas, but kept out of fragile regions known to accommodate rare and endangered species of flora and fauna.

To its advantage, the golden eagle is a very hardy animal that withdrew into the roughest and often most inaccessible regions of the mountains. It is in this rugged land of the golden eagle where intact mountain ecosystems still exist, enabling the king of the skies to thrive. On my excursions to the golden eagles, I am often deeply moved by the incredible beauty and diversity of flora and fauna which one might not have expected to exist anymore in the densely populated and highly exploited Swiss mountains. Regularly I encounter alpine meadows full of wild orchids or rare fire lilies that grow in fields with dozens of other rare blooming plant species. In such places, hundreds of different insects, from the tiniest beetle to the largest dovetail butterfly, visit the blossoms and altogether produce an incredibly intense humming sound which echoes from the surrounding rock walls. Bird species that I last saw in the 1970s when I was a small child, and that have disappeared almost everywhere, are common and dart around as if nothing has changed. Mammals, ranging from foxes, hare, ermine, and marmot, up to the large ungulates such as chamois or ibex, can be sighted easily. There is still a true mountain paradise left in Switzerland! Precisely this is the golden eagle habitat. It makes *A. chrysaetos* a flagship species of intact open and half-open landscapes of the Alps. The raptor's high ecological demands represent a very valuable indicator for the quality and biodiversity of this type of mountain terrain. Where golden eagles thrive, the subalpine and alpine ecosystems can be considered as near-natural. But golden eagles are the apex predators of those ecosystems and are therefore highly vulnerable if the quality and biodiversity of their habitats deteriorate. They will be among the first species to disappear!

After intensive research on the species and its biotope, ornithologists and ecologists consider *A. chrysaetos* as a potentially endangered or vulnerable species and issued several guidelines for protection.[68, 89, 94, 98] First, they warn that disturbance of nesting sites and their surroundings, as well as of hunting grounds, can put the species in danger. It is therefore of utmost importance to protect both nesting sites and hunting grounds to ensure longtime survival of a healthy golden eagle population. Second, researchers emphasize that local thermal systems are vital for the raptor's territorial fitness. Places where these thermal updrafts develop therefore need special attention and protection. And third, they found that breeding success strongly depends on continuous incubation of the eggs, undisturbed upbringing of the chick in the first 3–5 weeks, and the availability of prey in the breeding season from May to August. Hence, disturbance around the eyrie, and of game in hunting areas, should be kept at an absolute minimum. If we all work together and adhere to these guidelines, there is a high chance that one of the most iconic species of the Alps, the majestic golden eagle, will long continue to soar along vertiginous cliffs and snowy peaks touching hearts and souls of lucky observers.

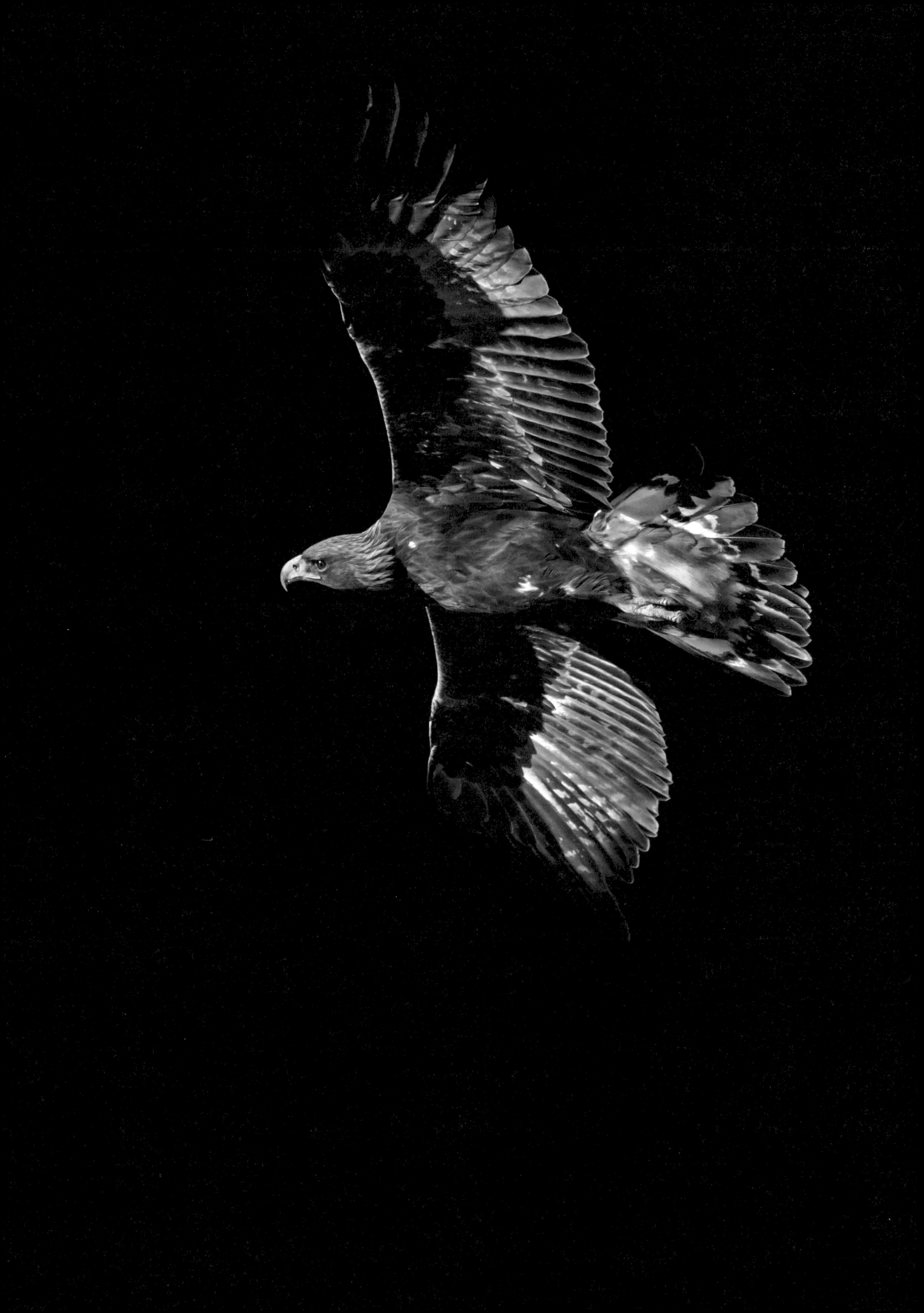

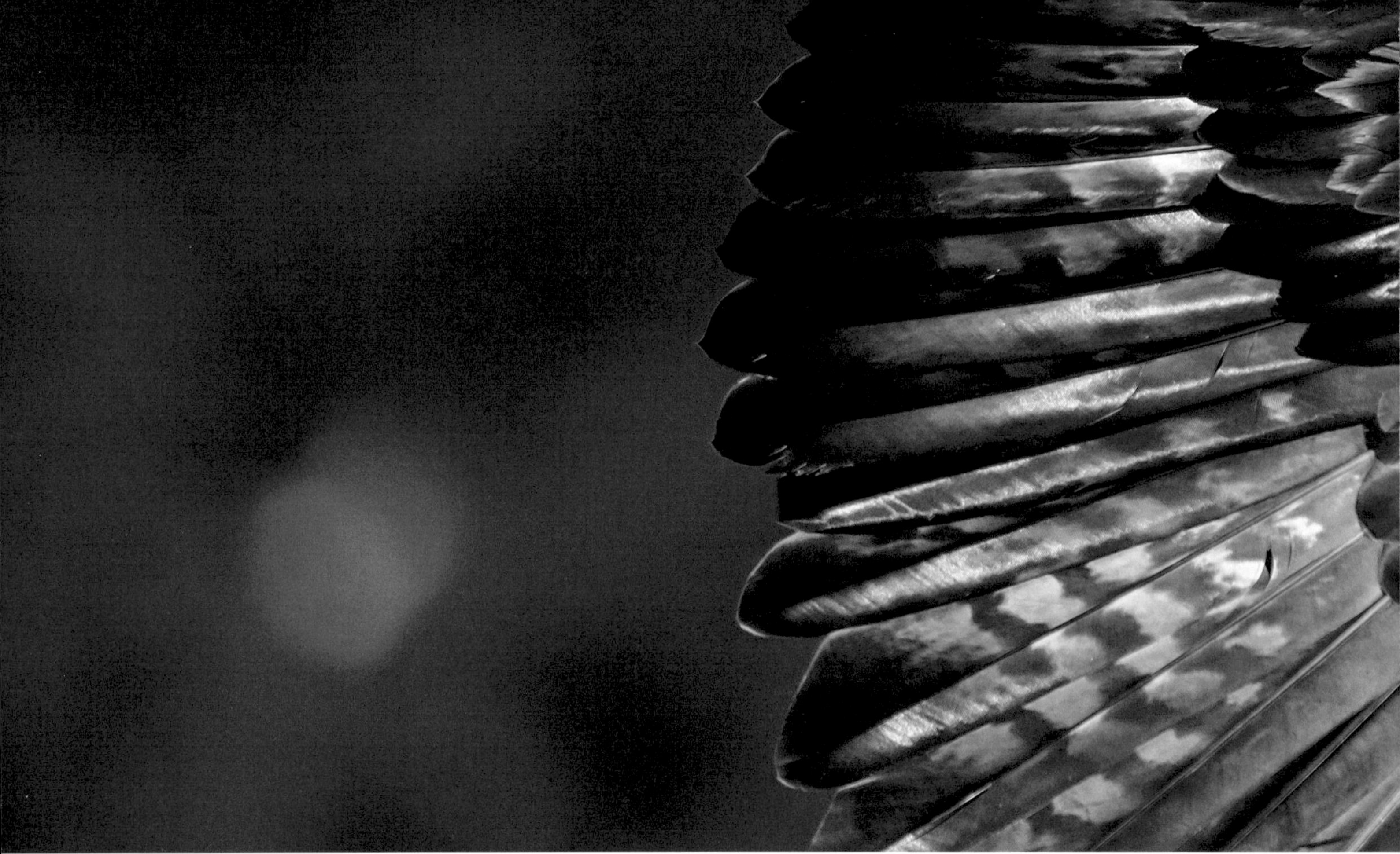

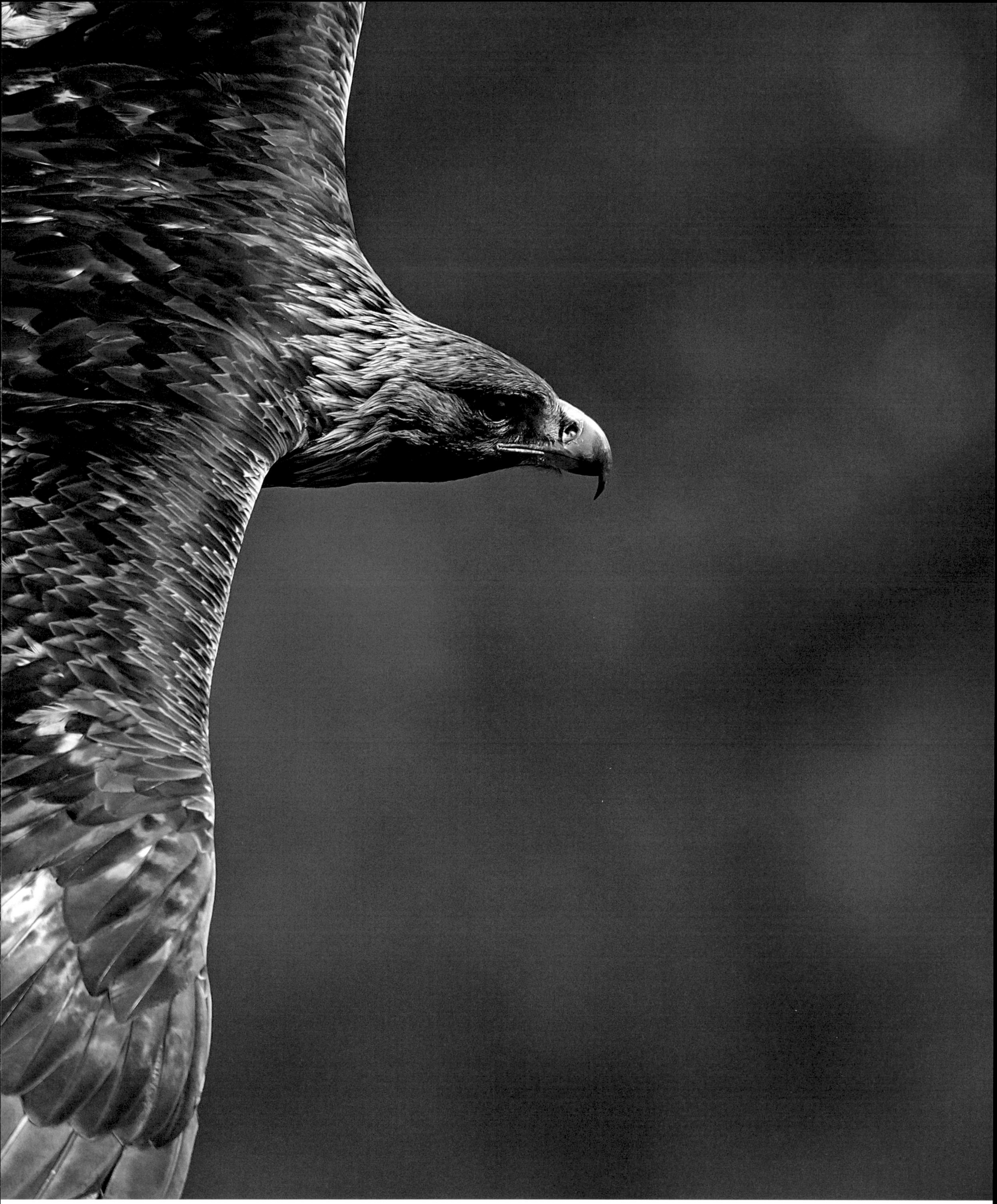

NO ONE SHALL CAGE AN EAGLE

WILLIAM BLAKE \
1757 — 1827 \ 99 \

When thou seest an eagle,
thou seest a portion of genius;
lift up thy head!

In other words, do not dare to think you can cage an eagle. It cannot be done; for an eagle caged is but divine energy prostituted to the tyranny of man; it ceases to be a portion of genius and is become a product of constraint, and a lie to the living truth. It is life robbed of purpose. \ 99 \

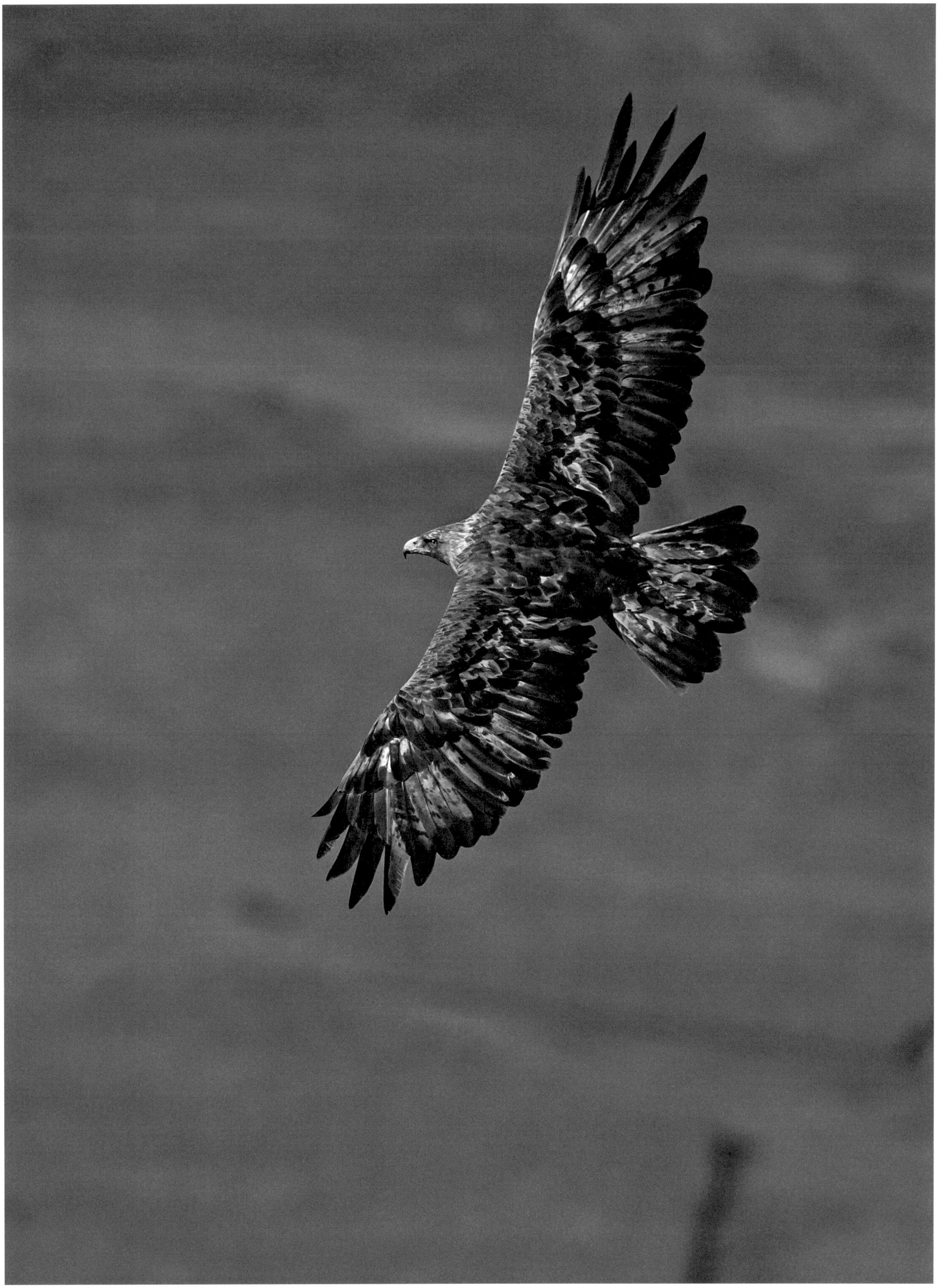

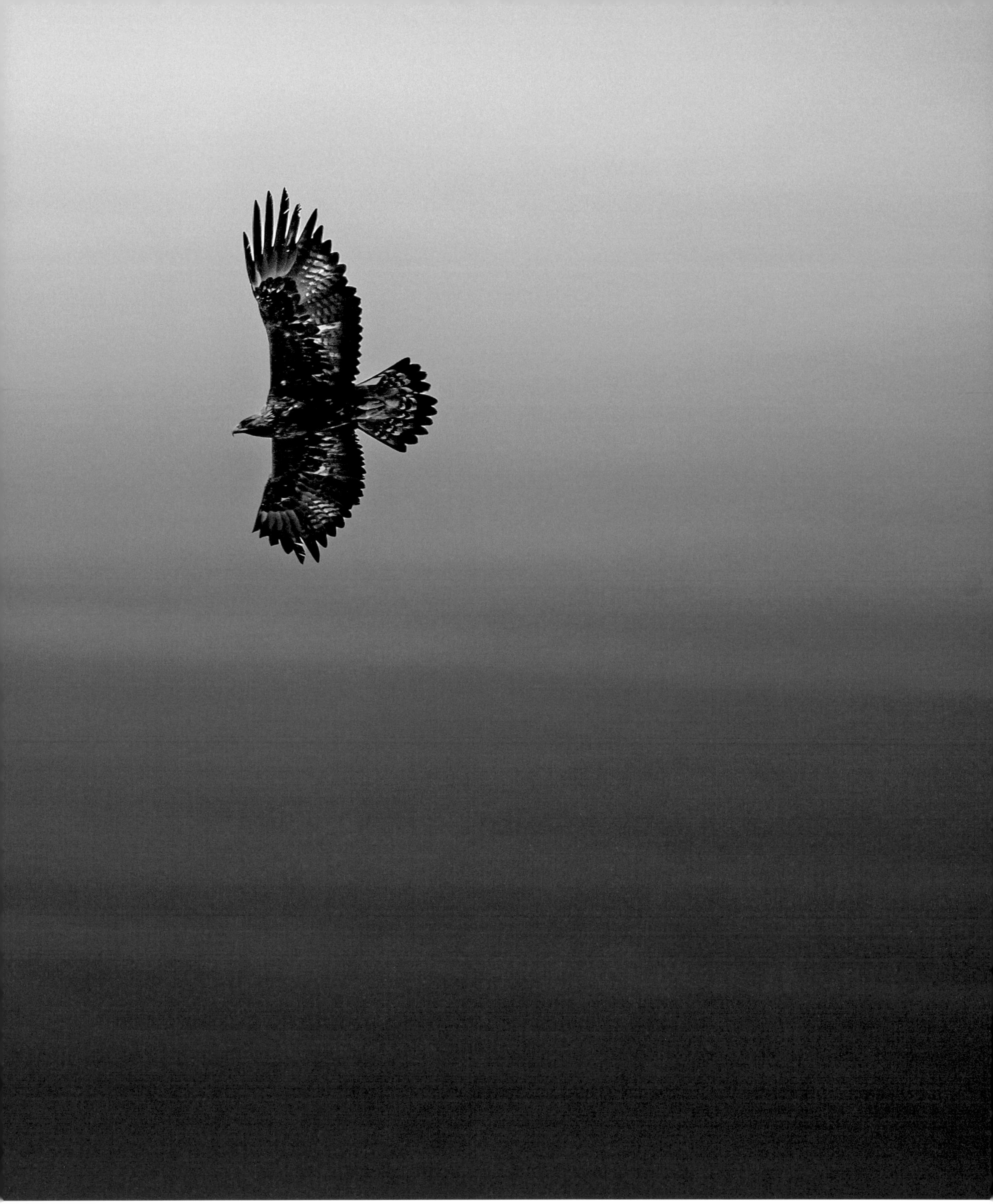

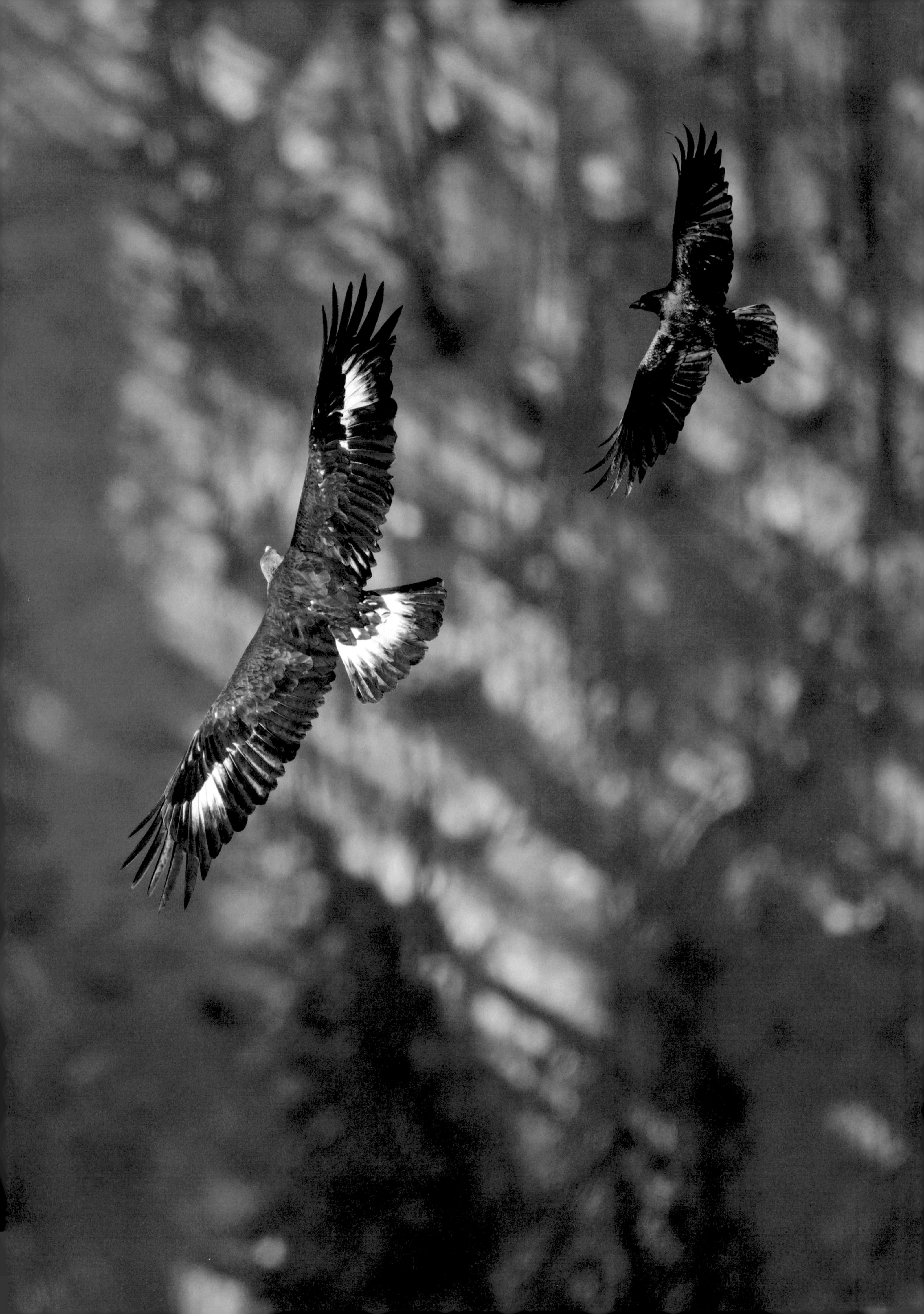

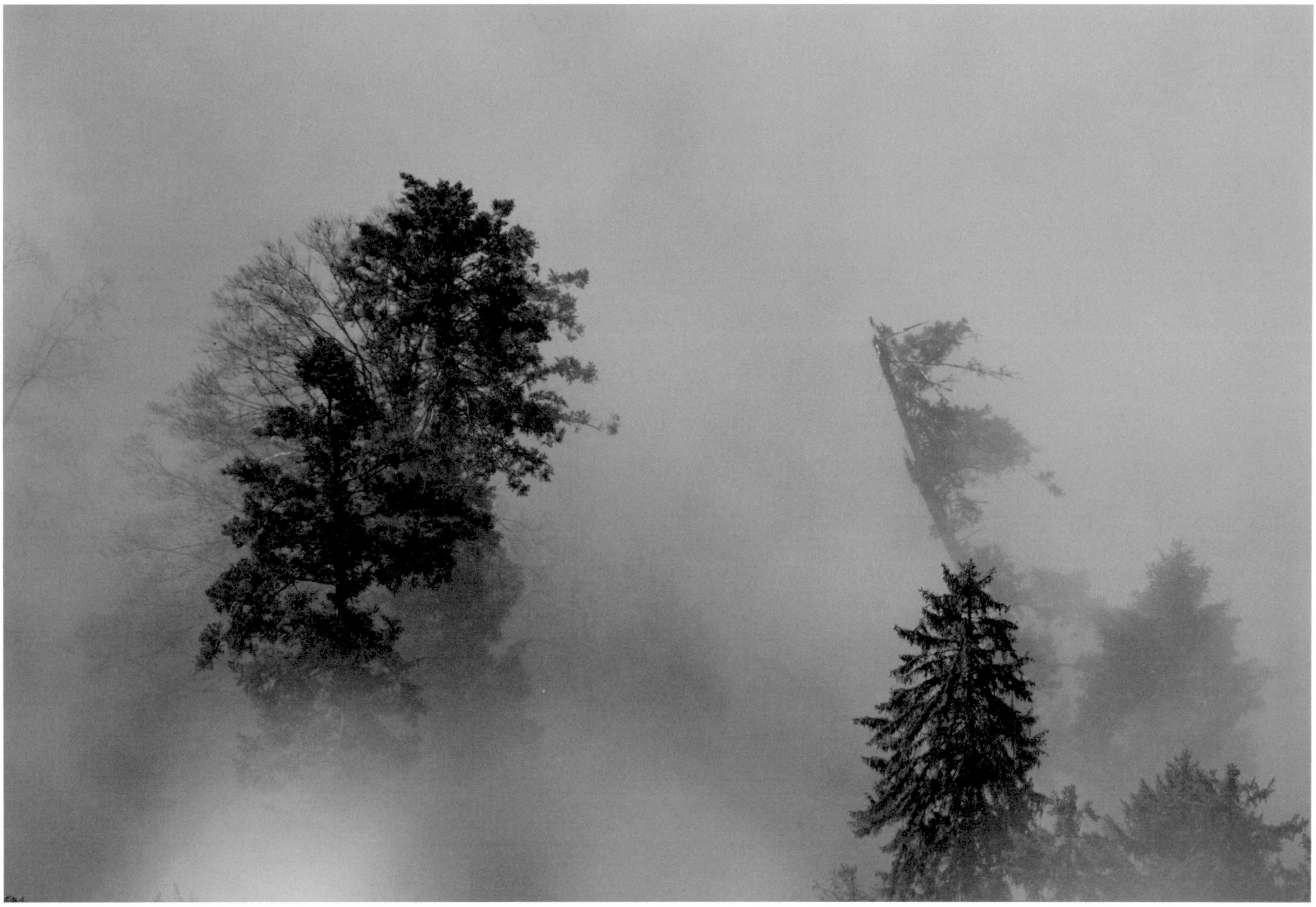

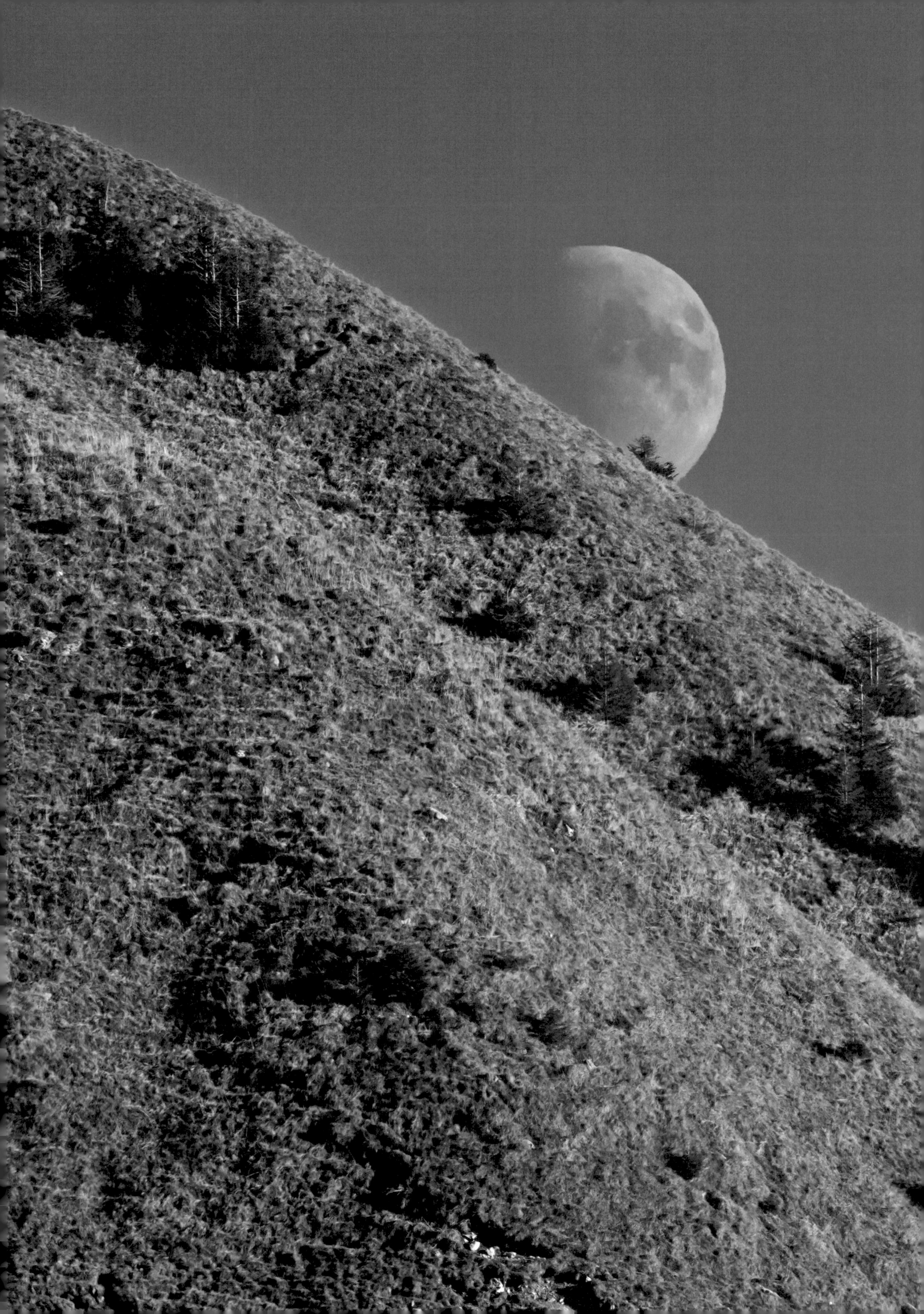

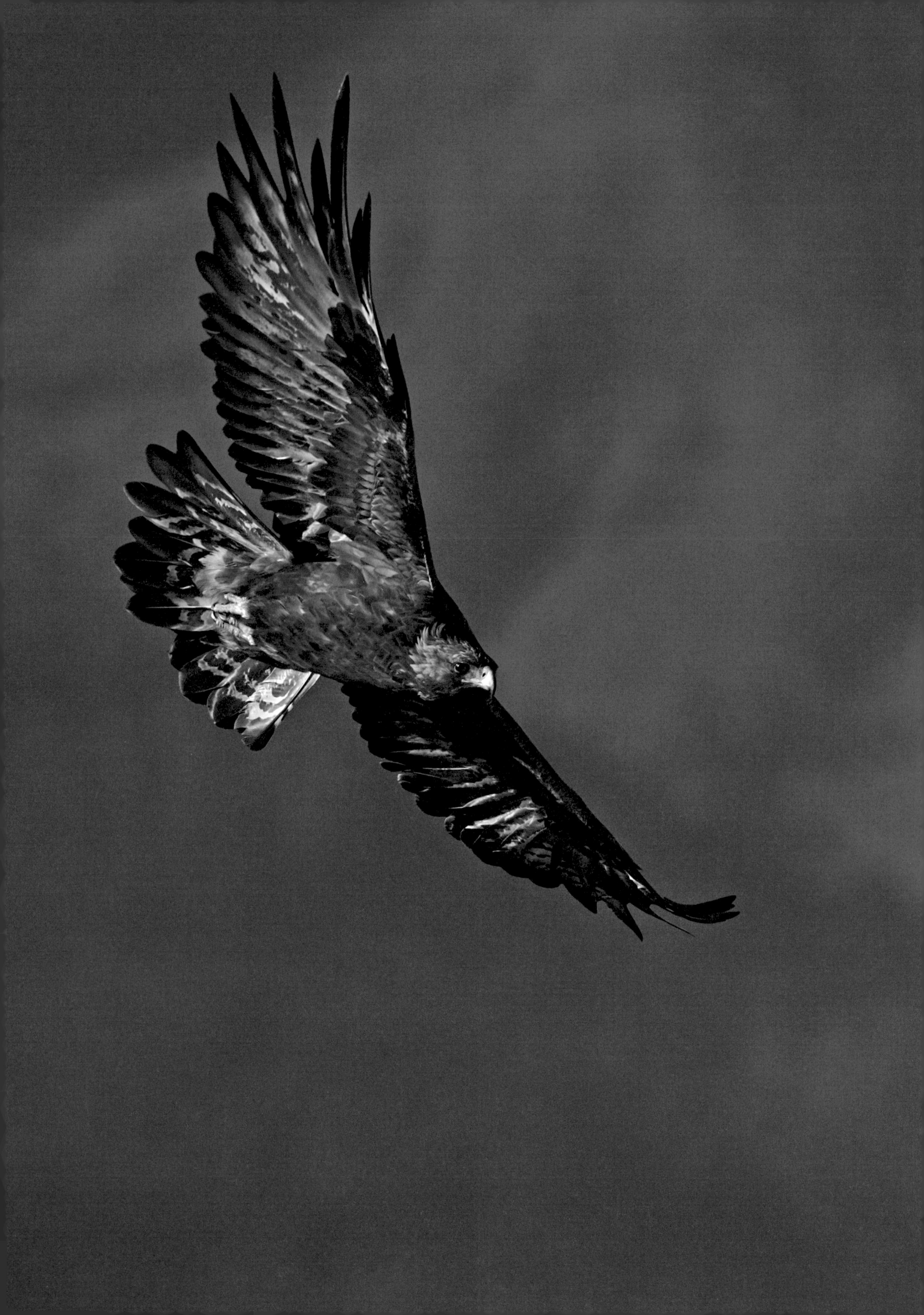

AQUILA, THE EAGLE GIRL

A FAIRY TALE FROM THE TICINO IN SWITZERLAND. TRANSLATED AND SLIGHTLY MODIFIED \ 100 \

On a warm summer day in July, a young widow, I don't remember from which village, hiked up the mountain to make hay. She carried a large basket on her back with her two-year-old child sitting inside. The little girl was a real sweetheart, cute, and lovely. While her mother was cutting fresh grass with a scythe, the girl happily ran around on the alpine meadow and collected wild mountain flowers. But suddenly, a mighty golden eagle stooped towards the child like a falling star, snatched her with its sharp talons, carried her high into the sky, and flew back with her to the nest. Imagine the shock, the desperation, and crying of the distressed mother. But the strange fact was that the child was not at all scared of the powerful raptor. She snuggled happily up to its golden neck, giggled, and twitched its feathers. The huge eagle liked the girl's innocent caresses and started to feel affection for her. The fearsome predator decided to take care of the little girl as if she were his own daughter. He flew out from the nest and brought back wild honey and sweet fruits to feed the girl. He taught her how to hold on to the steep cliff and how to climb tall mountains.

One day, the eagle decided to fly deep down into the valley to visit several villages where he stole clothes and little dresses that the farmer girls had hung on washing lines to dry in the sun. Later, when the girl grew bigger, the eagle wanted nicer clothes made of velvet and silk for the girl. Hence, he flew far over the land to palaces and castles of queens and princesses. He stole their magnificent dresses and jewelry and carried them high up into the sky to the inaccessible heights of his nest. One queen, from whom the eagle had stolen particularly beautiful items, got very angry and asked her son to hunt down the thievish raptor. The prince replied to his mother: "I don't want to do what you tell me." But later, when the prince was alone, he thought: "Why in god's name does this bird steal all those dresses and jewelry? What can he use it for? I have to fathom this mystery." For several months, the prince roamed far distances in the mountains and climbed to the top of many peaks, always looking for the golden eagle. "I think I need to end my quest. I have been searching for many months now, but the curious eagle is nowhere to be found." It was a wonderful day in May and the prince just wanted to return home when he heard a sweet girl's voice singing high above him. Immediately, the prince climbed up the mountain and found the young singer, light-hearted in the large eagle's nest. She was the most beautiful girl he ever had seen. She reached over to take his hand and invited the prince to sit next to her. This is how they became friends and told each other their adventures.

The girl detailed for the prince her wonderful life with the golden eagle, in the marvelous and secluded mountains. And the young man explained that he was a prince and lived far away down in the valley. He asked her: "Would you like to become my wife and live in a palace?" The eagle girl joyfully agreed, and together they climbed down the cliff and then the mountain, finally to reach the king's castle where the prince presented the girl to his father: "Look father! I found this beautiful girl high up on a mountain in an eagle´s nest. We love each other and I would like to marry her." The king liked the girl and gave her a welcome kiss. "I will name you Aquila, which means eagle." He smiled with affection and instantly started preparing for a glamourous wedding ceremony.

But the mother of the prince, the queen, did not agree at all with his choice. "What are you thinking of? Marry a totally unknown girl? Who is this shabby girl from the mountains?" And she ordered the servants in secret to drown her in the river. They obeyed and threw the girl into the stream. But the mighty golden eagle heard the desperate screams of the drowning girl and dived down like a flash of lightning, grabbed her carefully with his sharp claws, and brought her to the riverbank. After she had recovered for a while, Aquila returned to the palace.

Soon a majestic wedding took place, where the handsome prince married his lovely bride. To punish the cruel queen, the king abdicated soon after and handed over the throne to his son, so that his beloved daughter-in-law, Aquila, became the new queen.

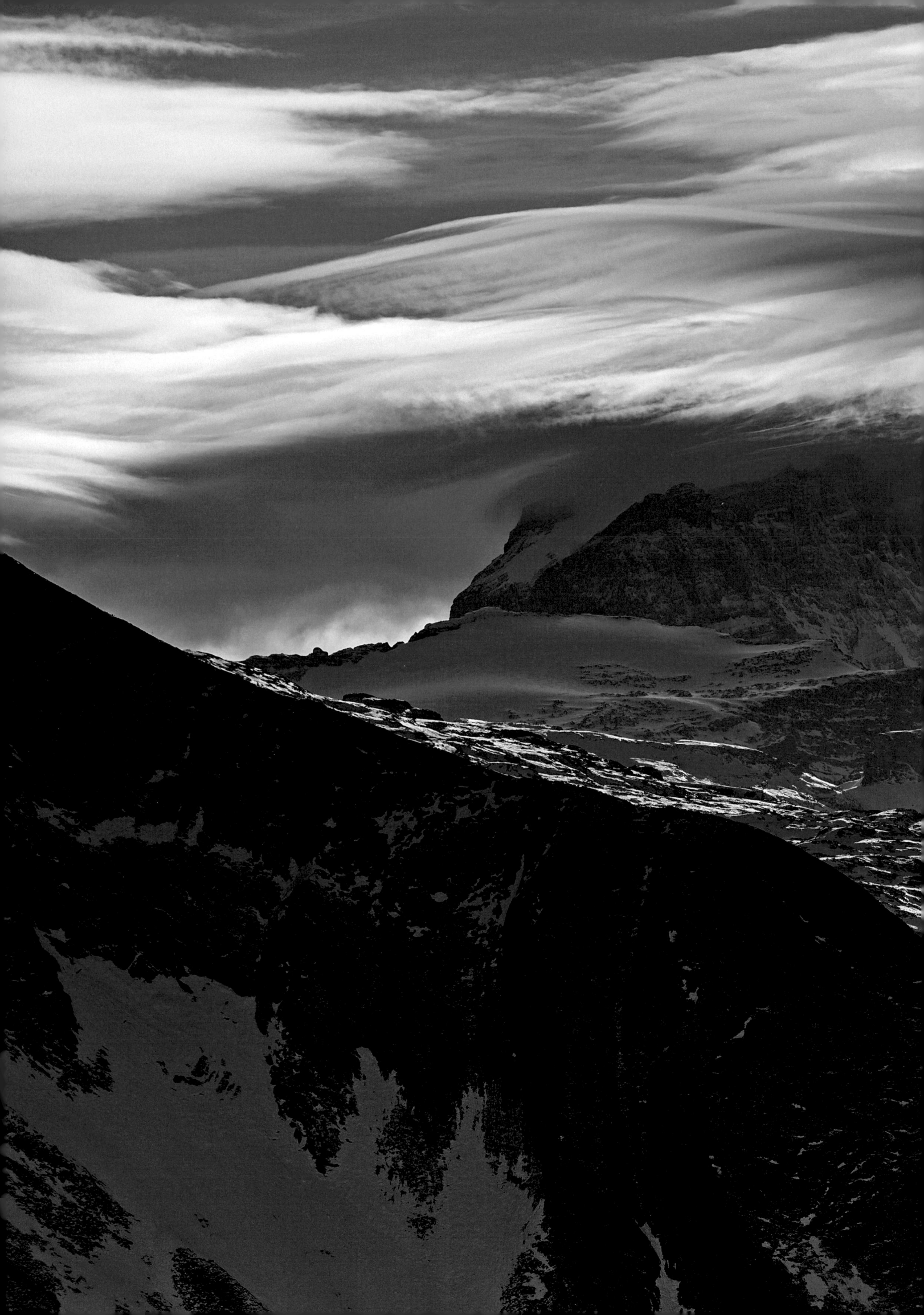

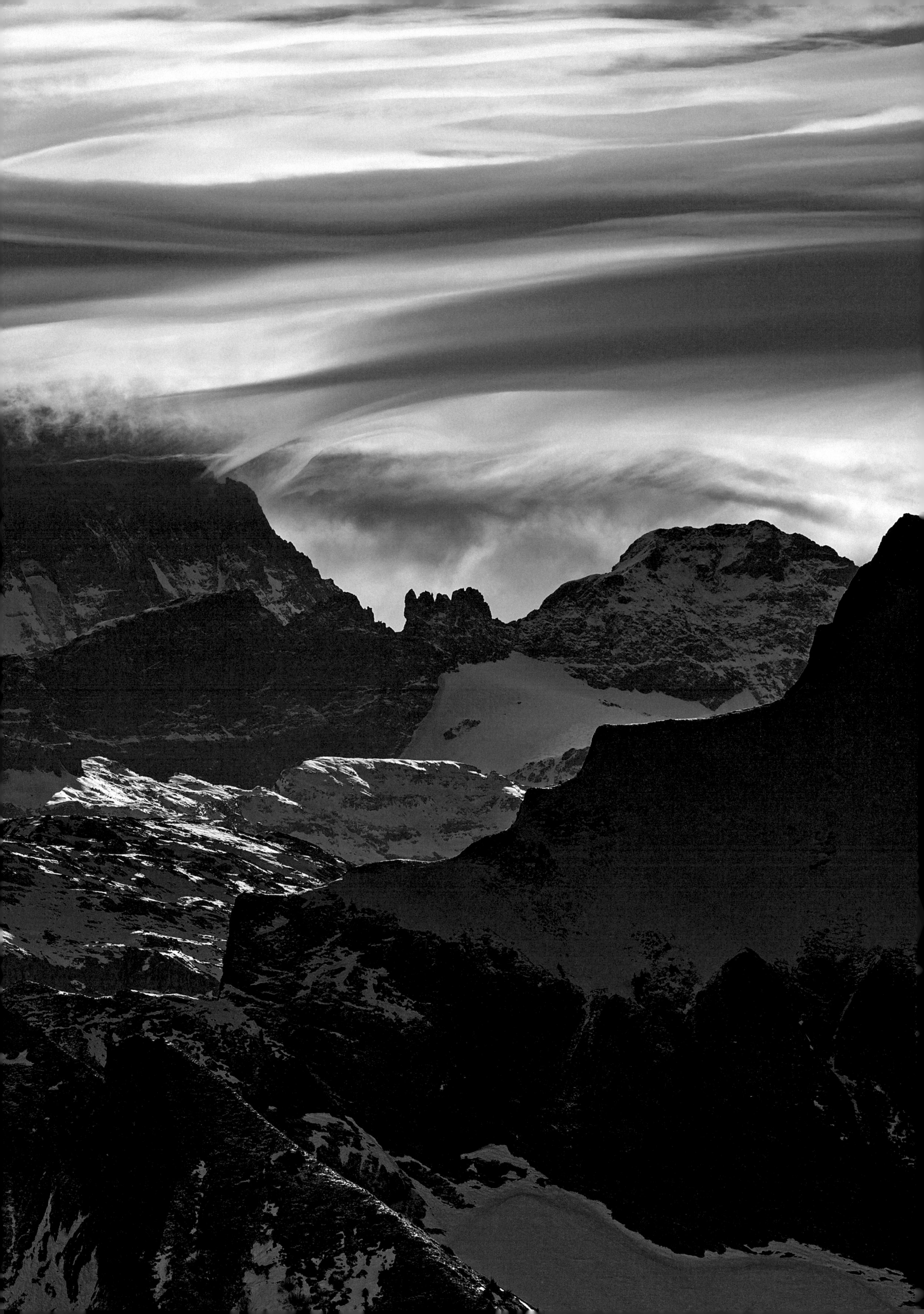

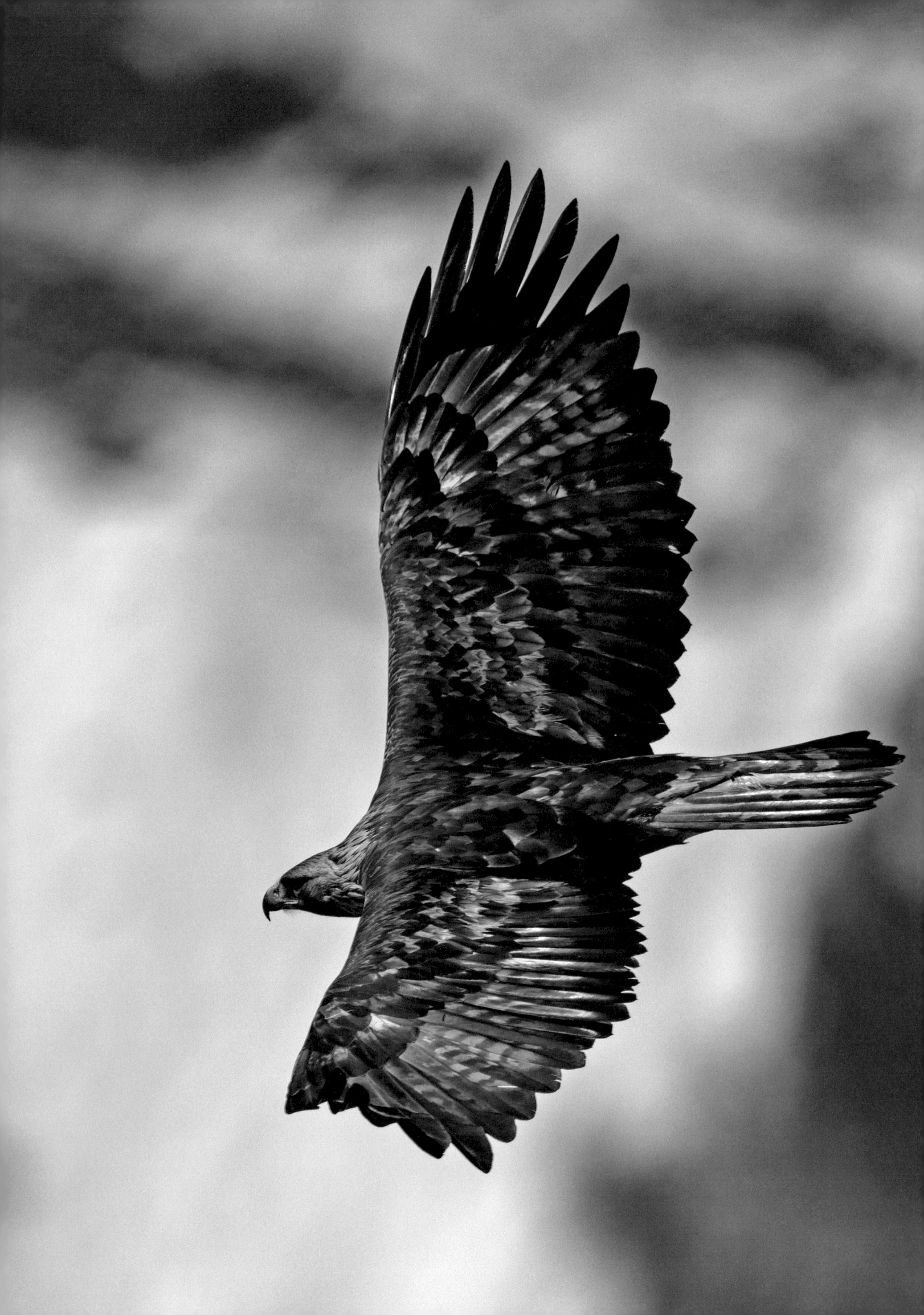

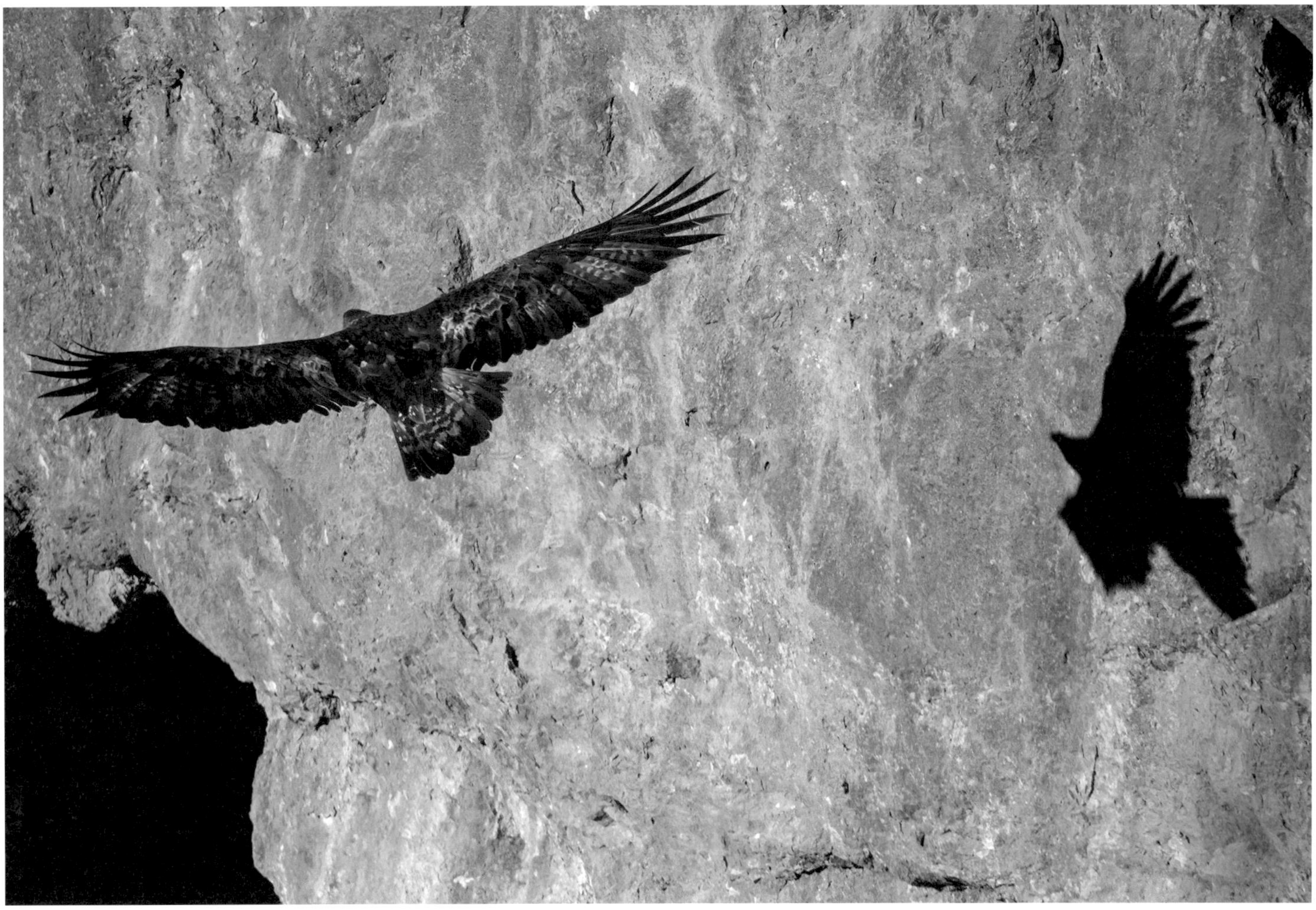

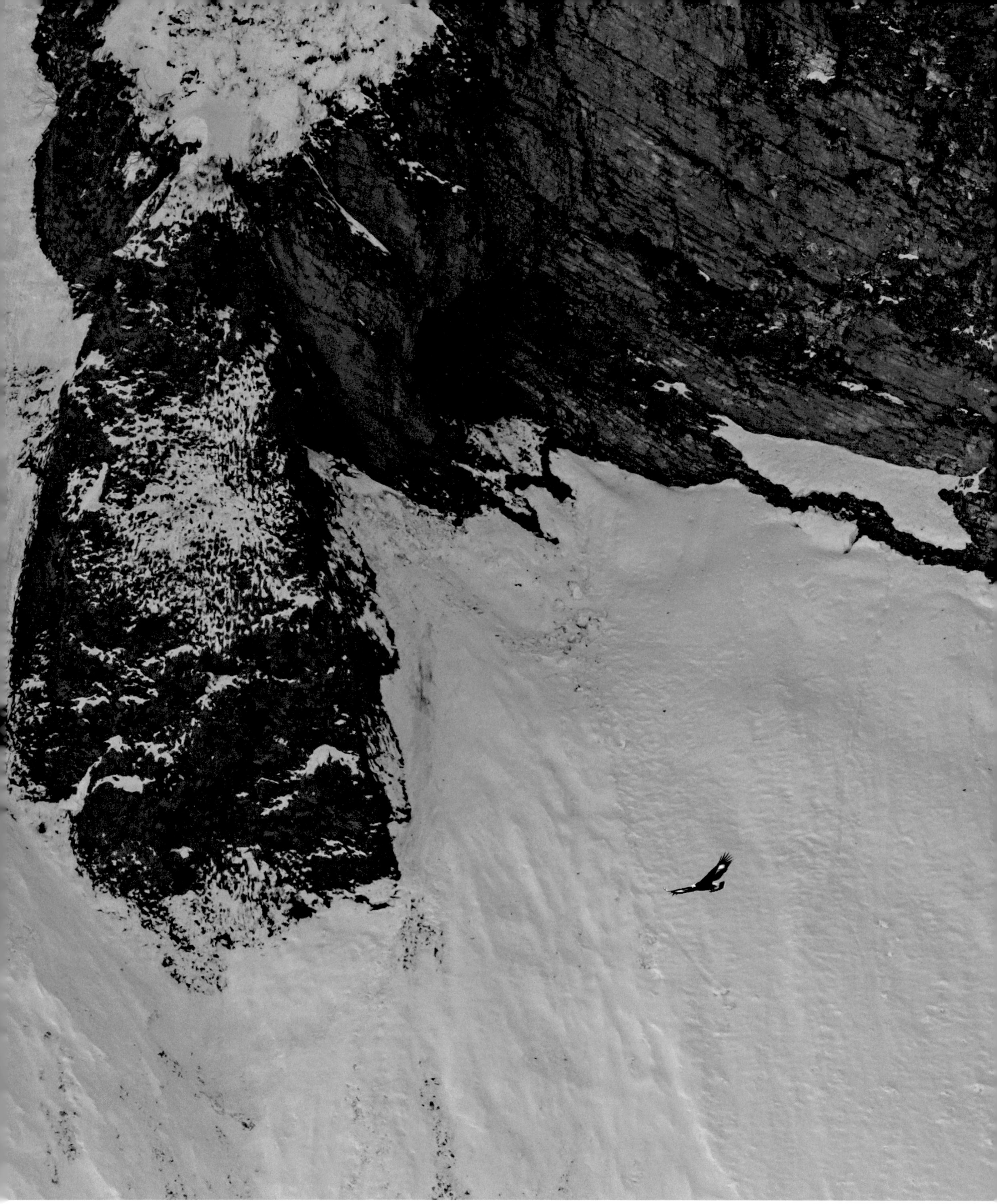

AXAYACATL'S ARMY MARCHES TO THE PUREPECHA

AXAYACATL, AZTEC POET
AND LORD OF TENOCHTITLAN \
CA. 1449 — 1481 \ 101 \

Now we are approaching Michoacan,
they have fallen on them,
they will certainly fall, the old Mexica warriors,
they will come to expose themselves to danger,
will come to finish the work, the old eagles,
the warrior,
the experienced eagle,
Huitznahuatl,
the ancient nobility.

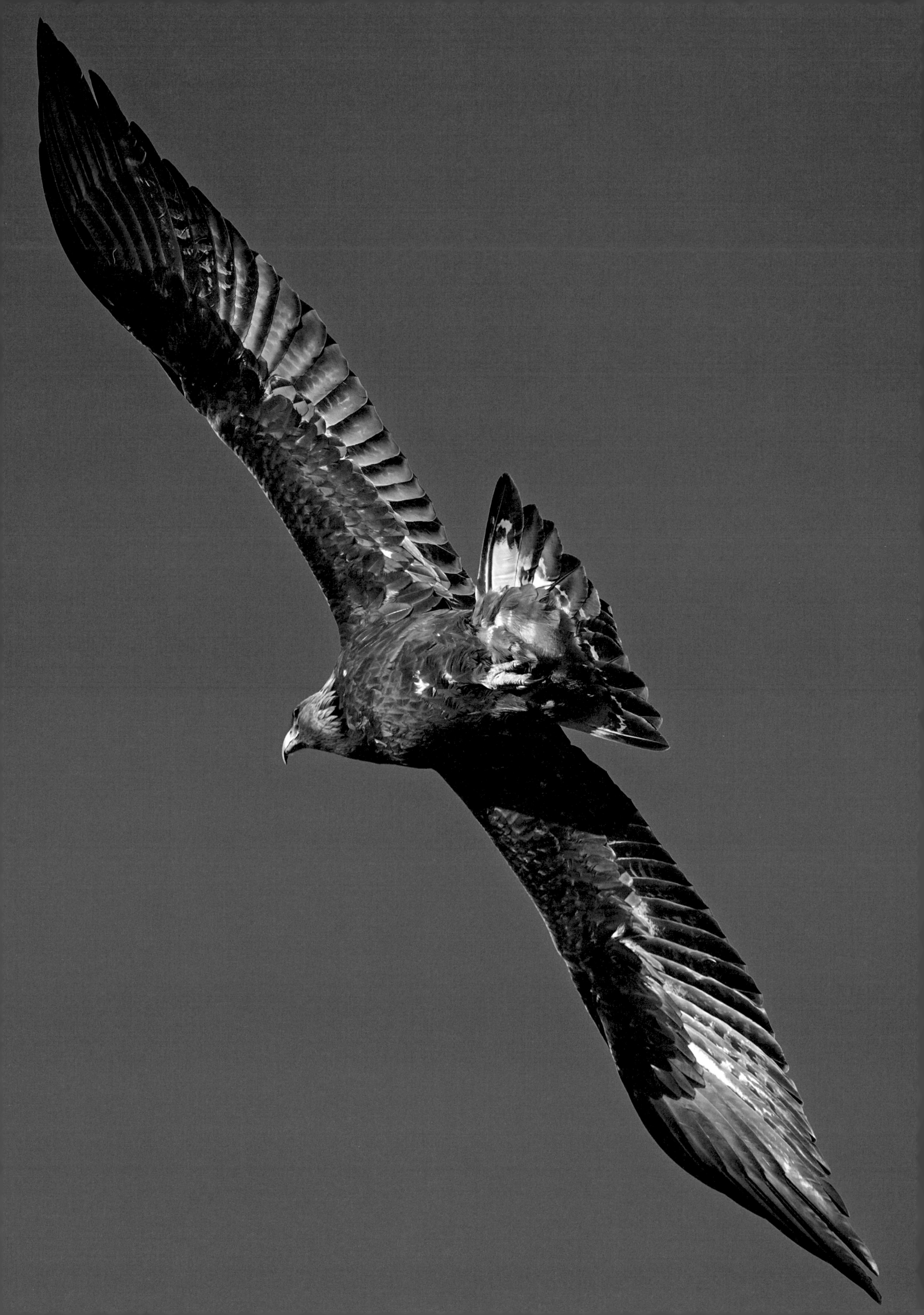

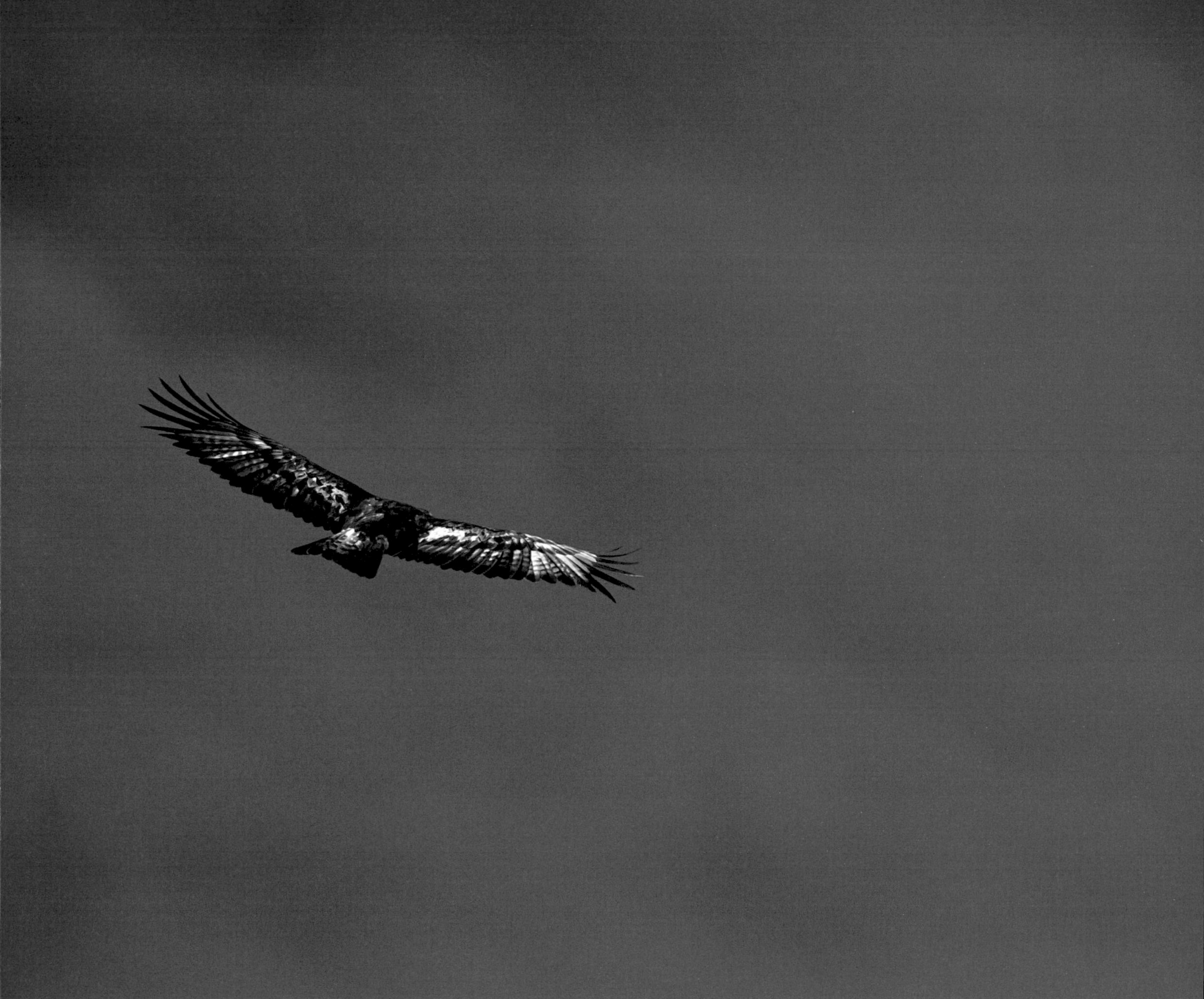

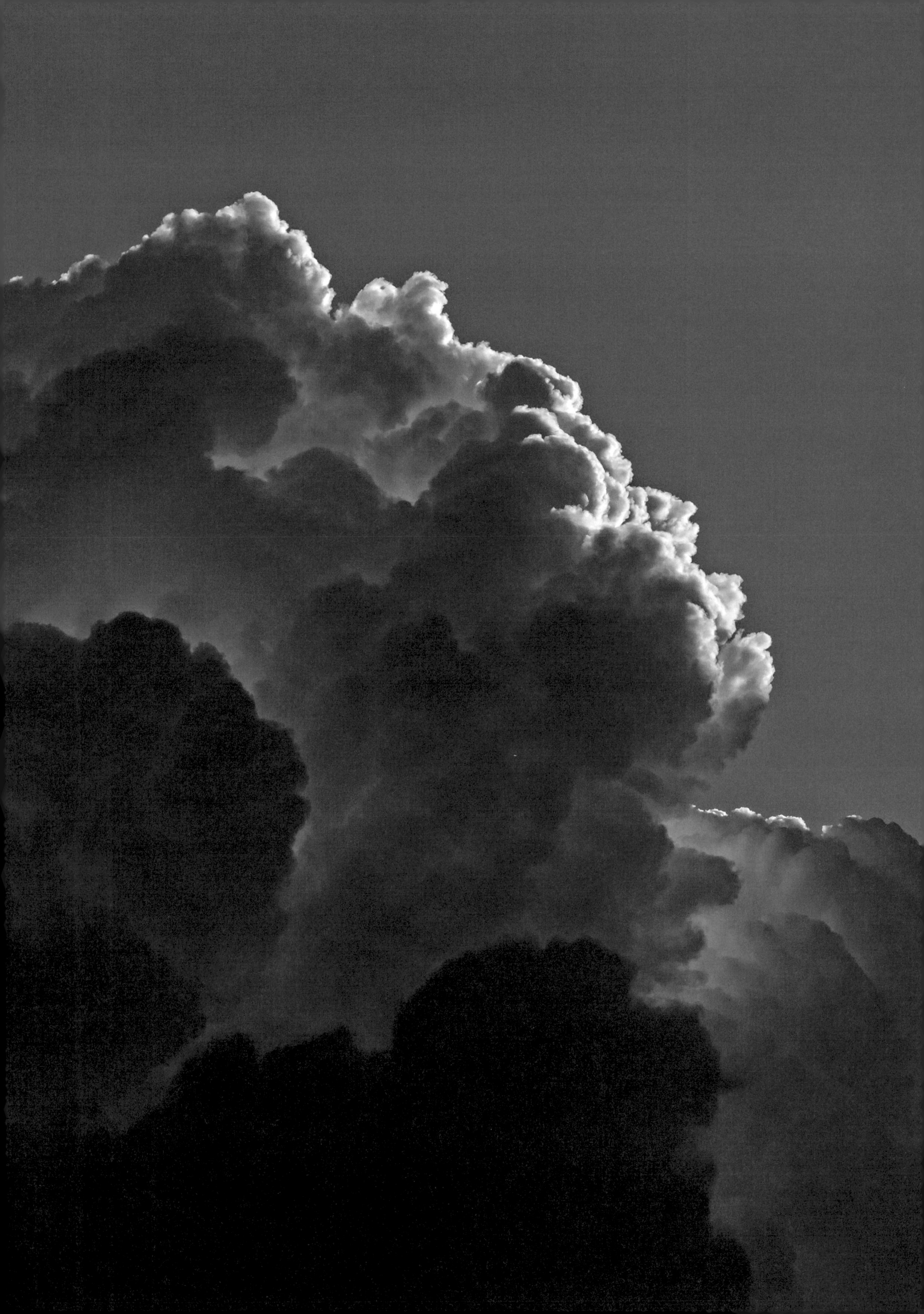

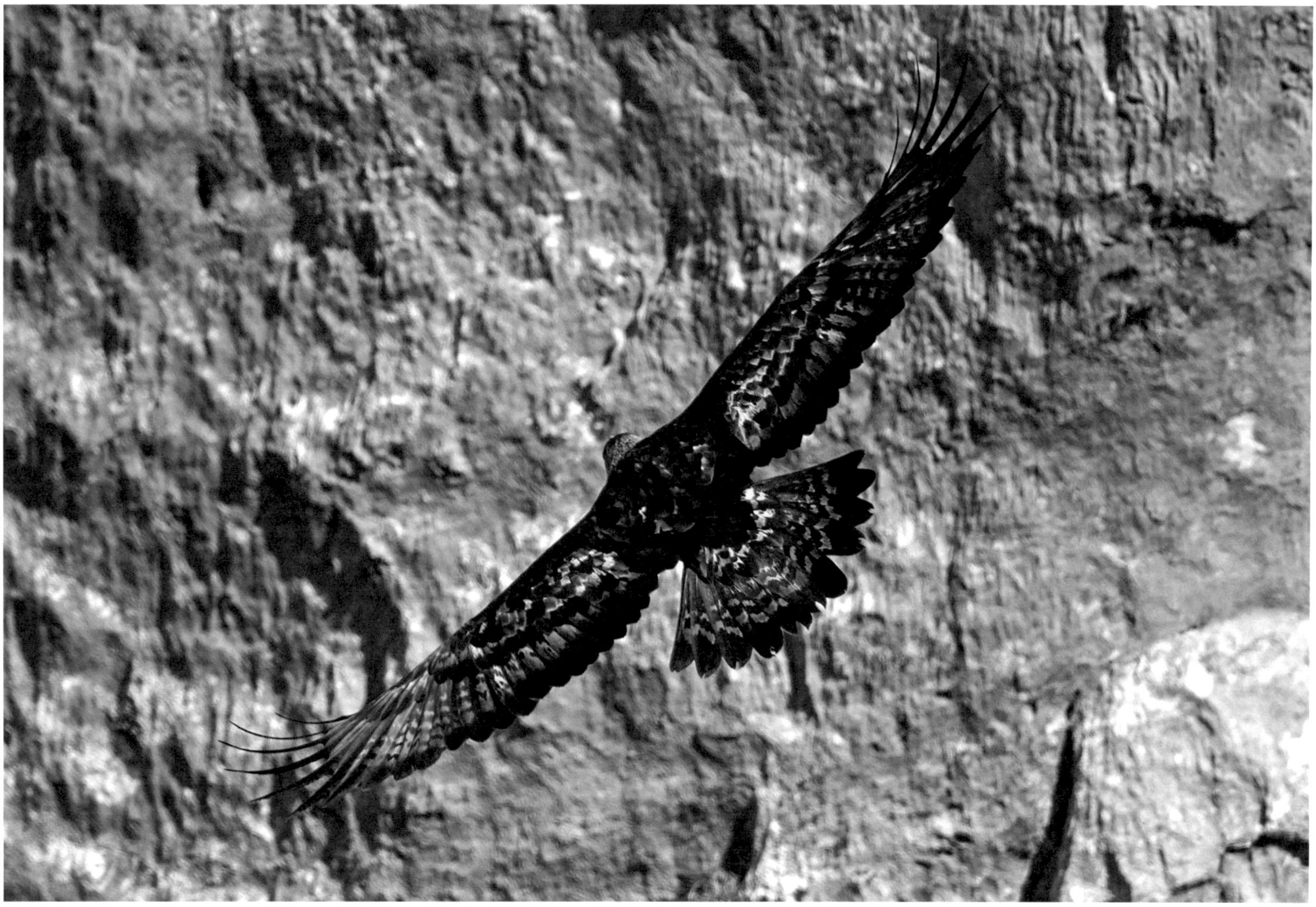

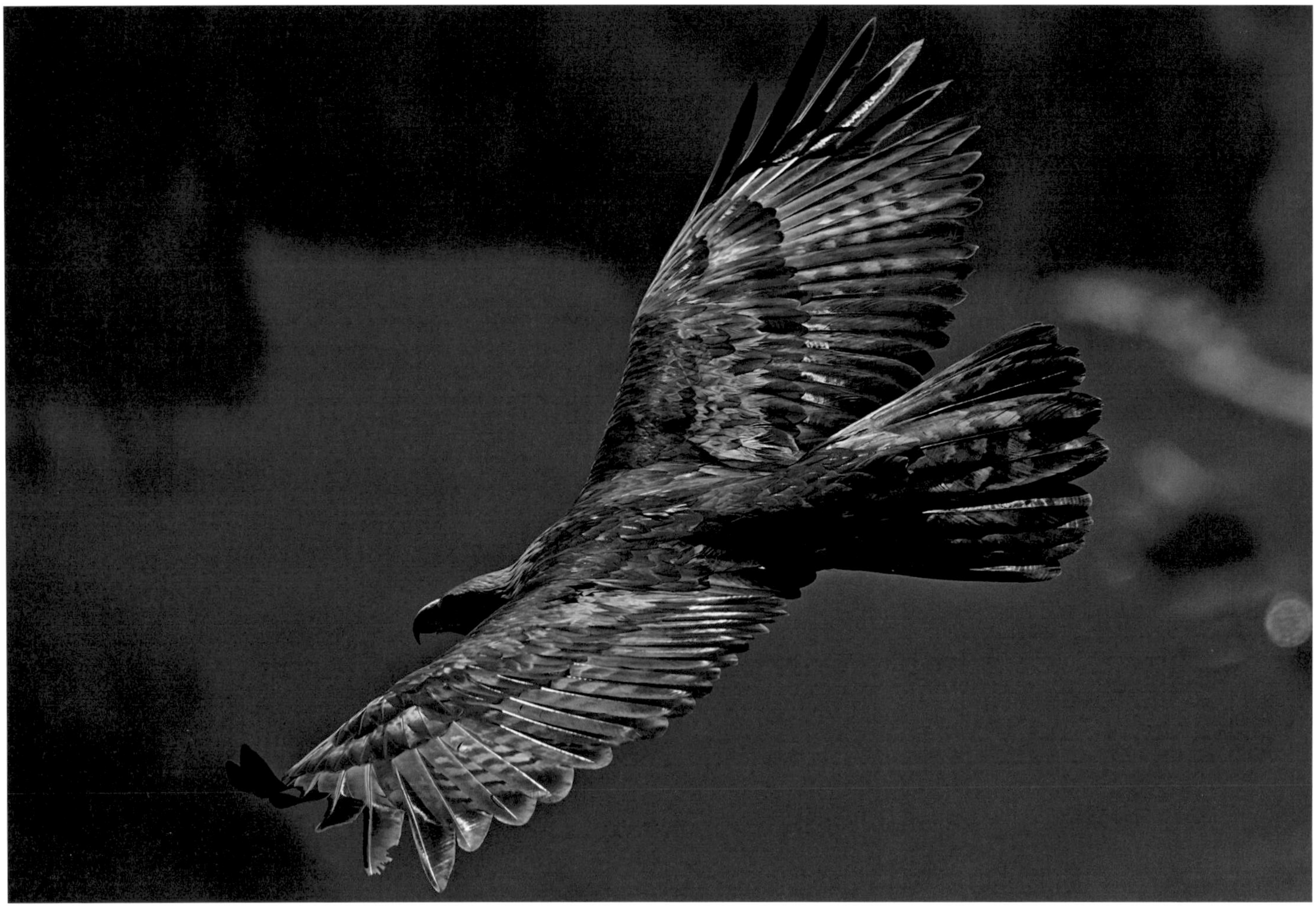

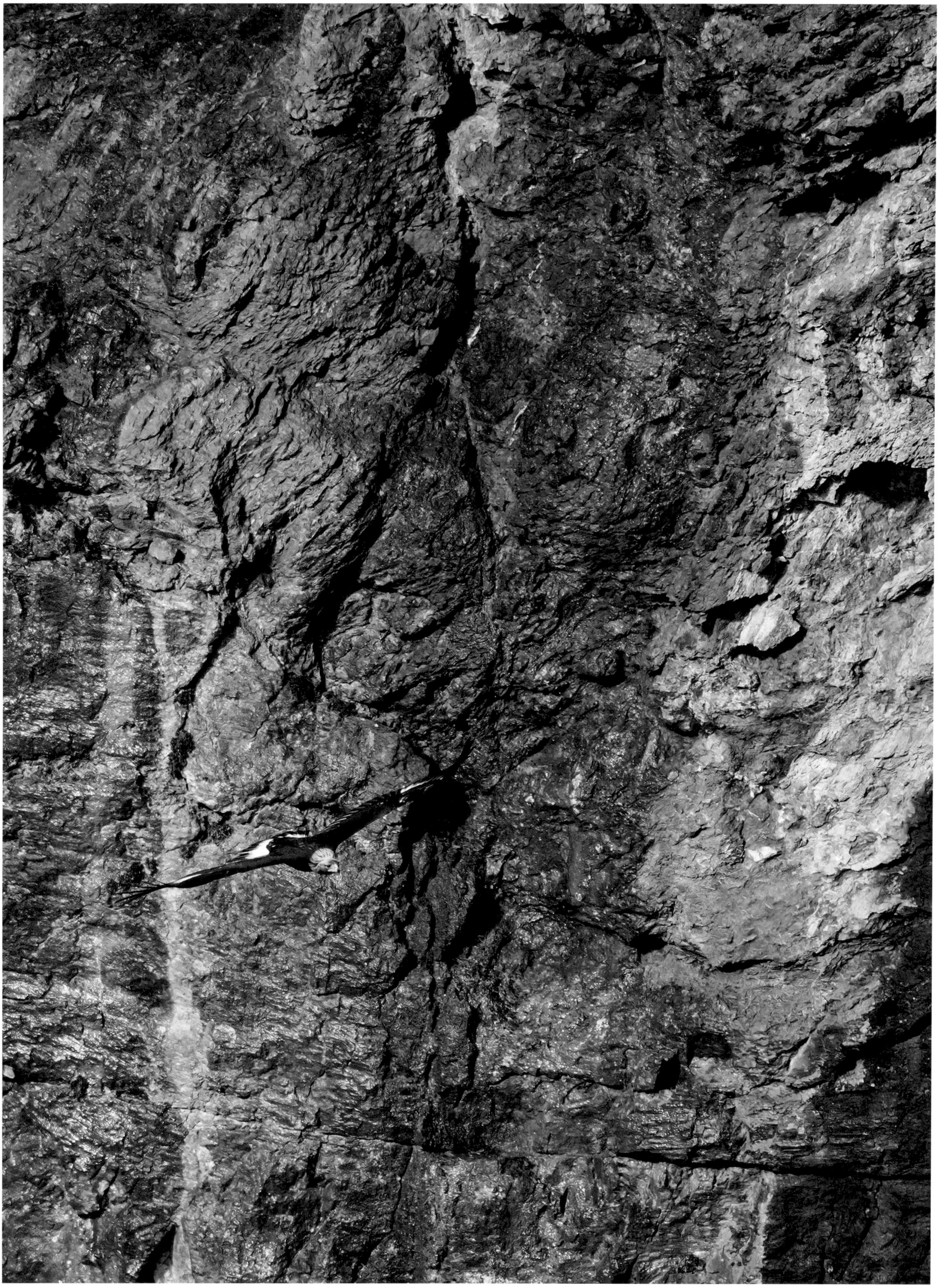

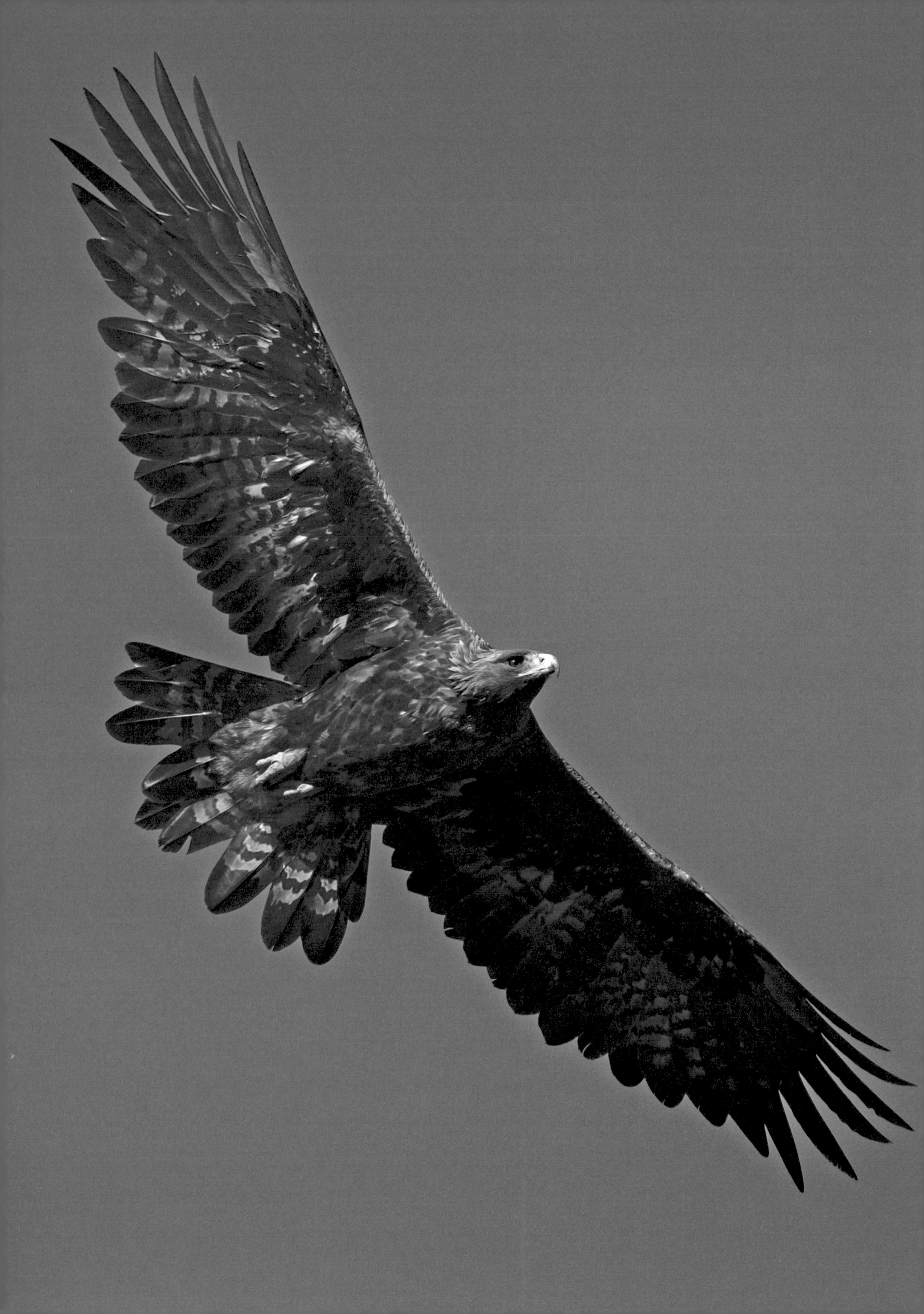

WITH THE DETERMINATION OF AN EAGLE

CRAZY HORSE
(TȞAŠÚŊKE WITKÓ IN LAKOTA) \
CA. 1840 — 1877 \ 102 \

A very good vision is needed for life,
and the man who has it must follow it –
as the eagle seeks the deepest blue of the sky.

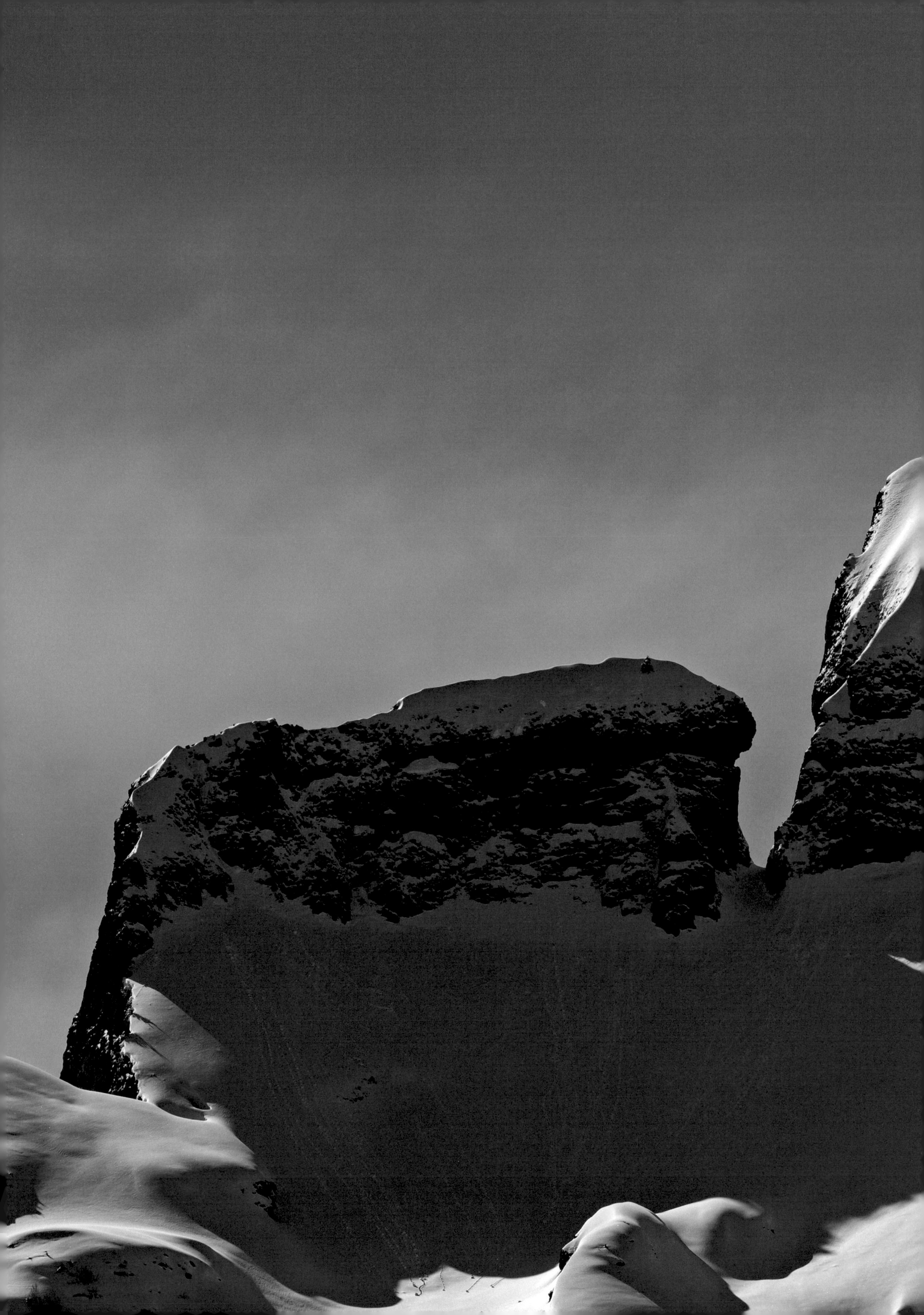

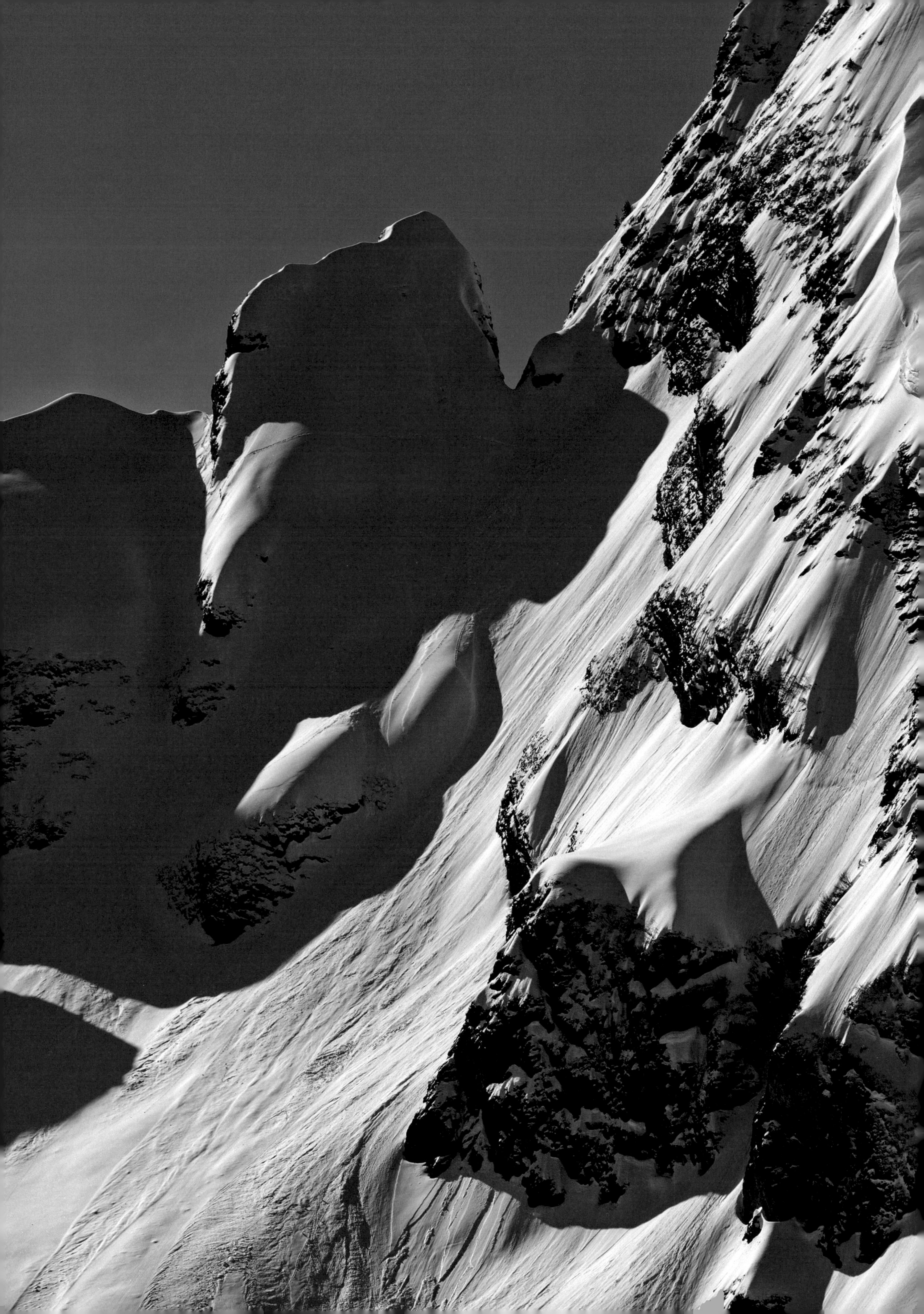

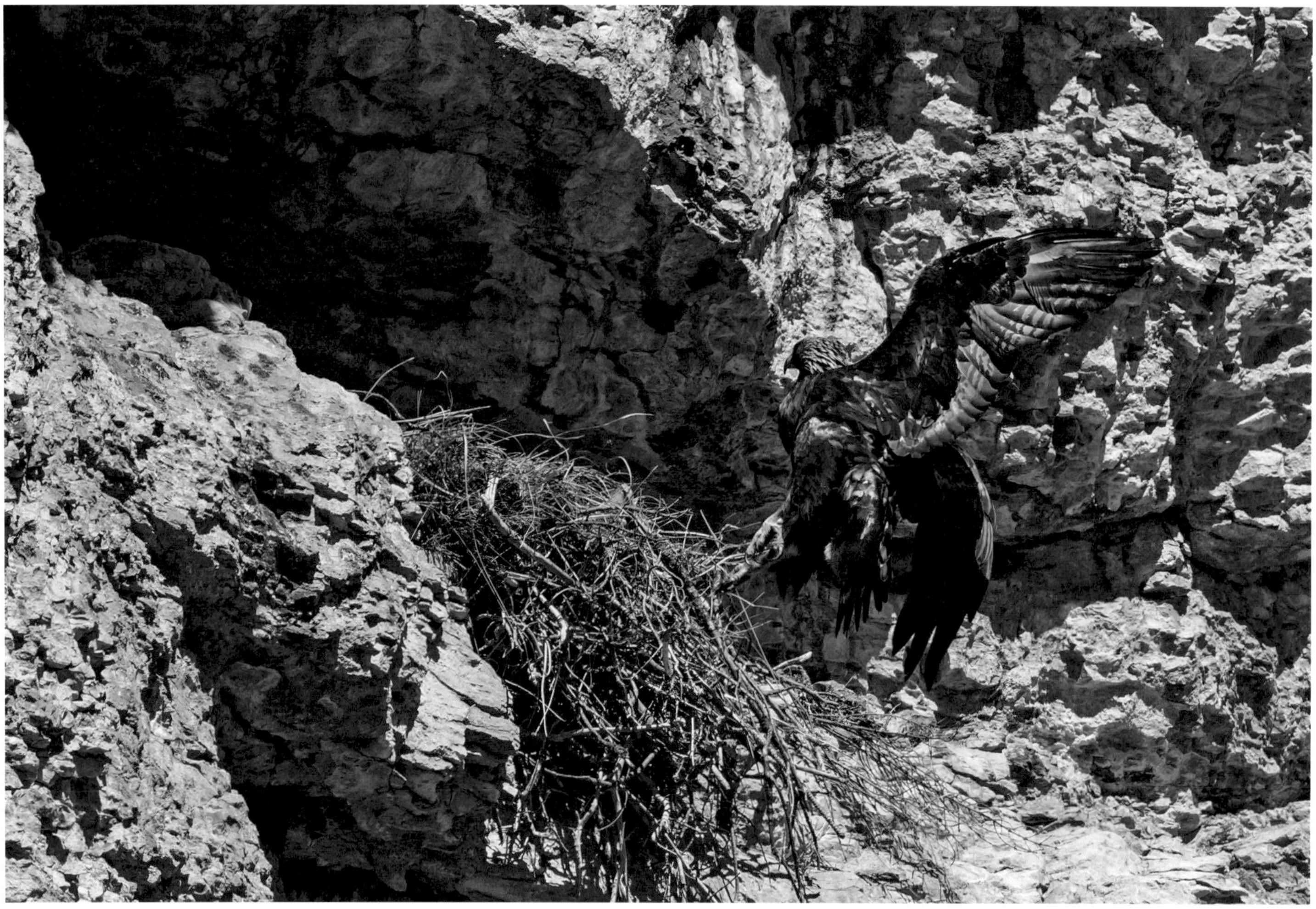

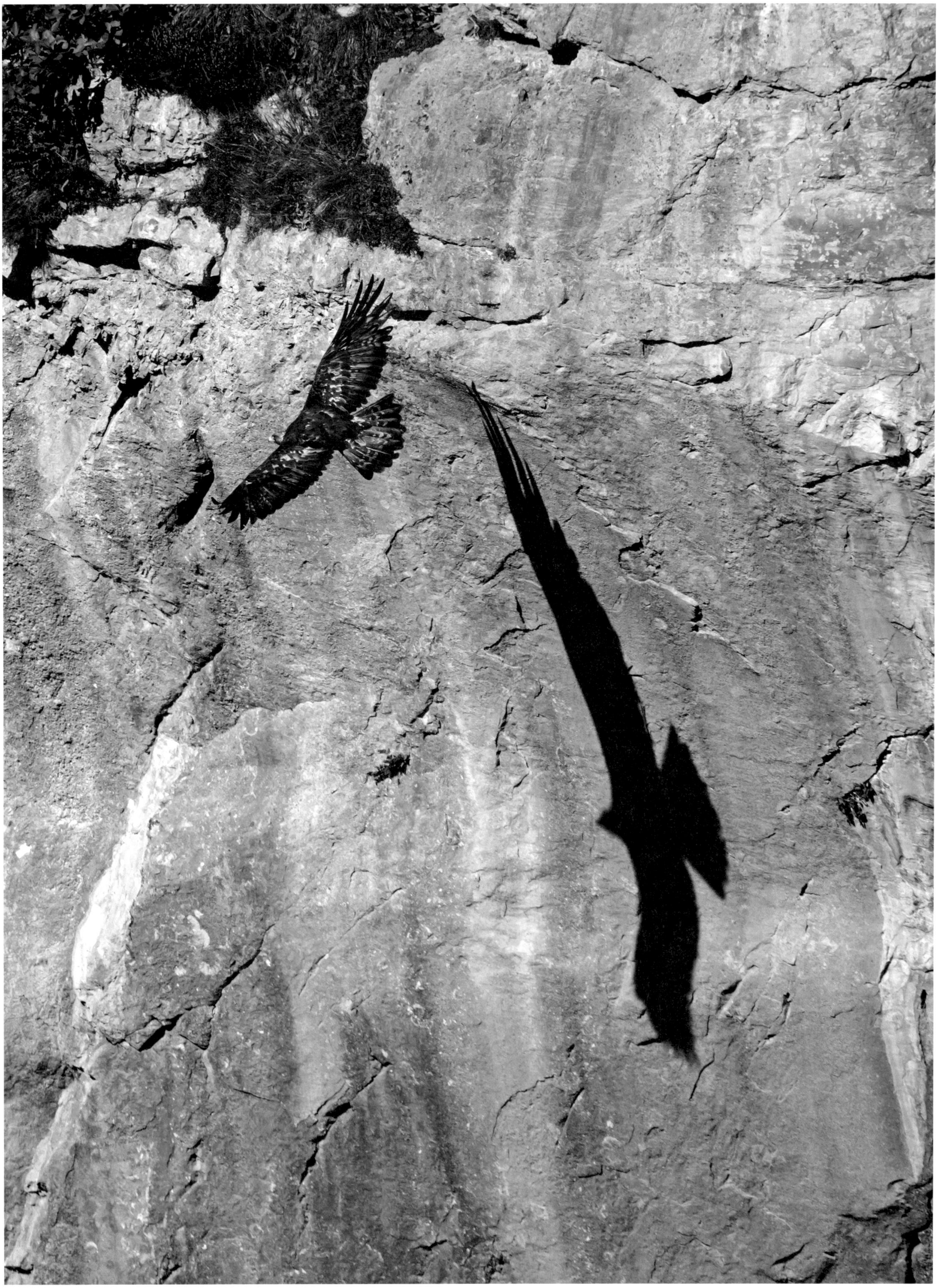

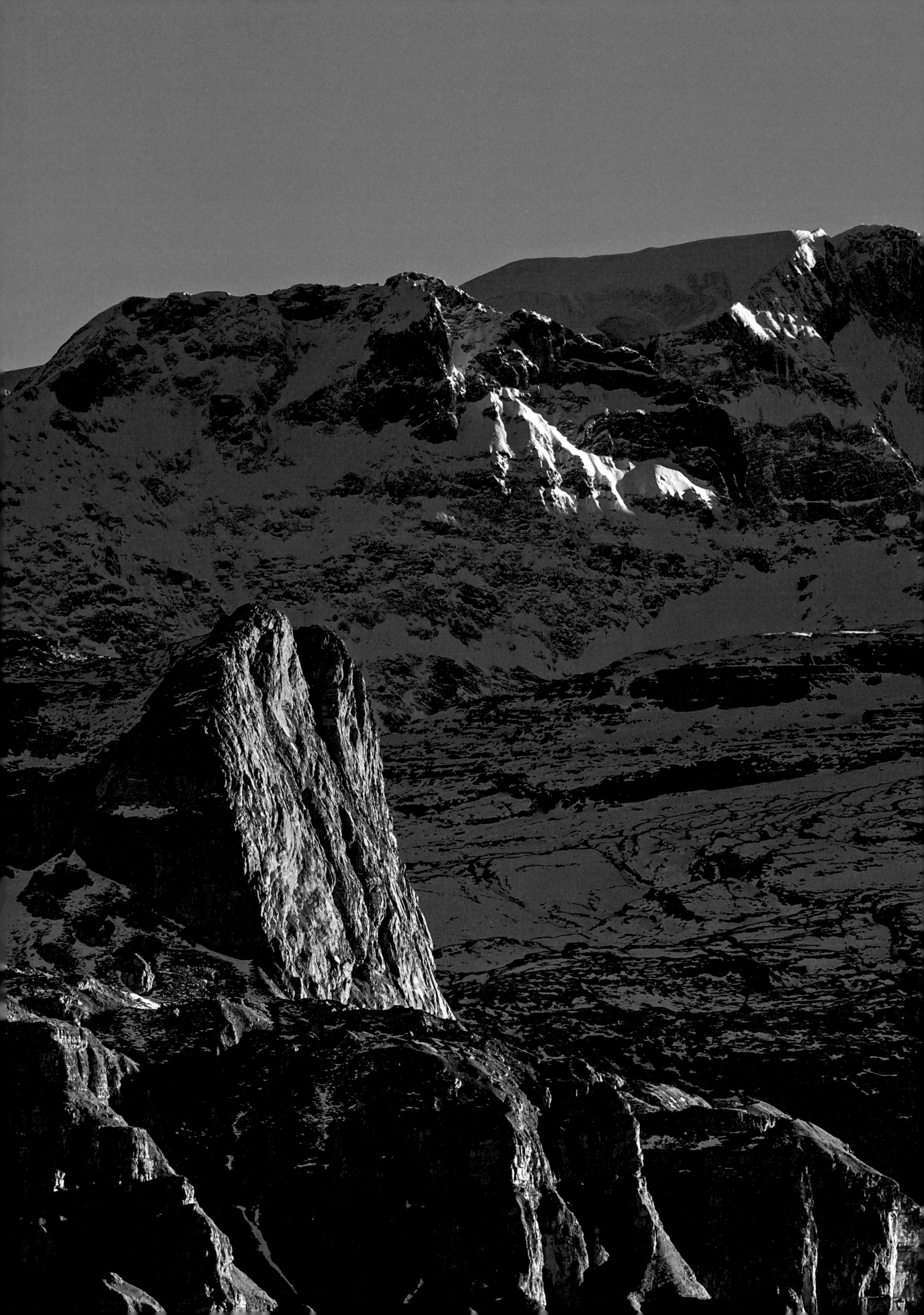

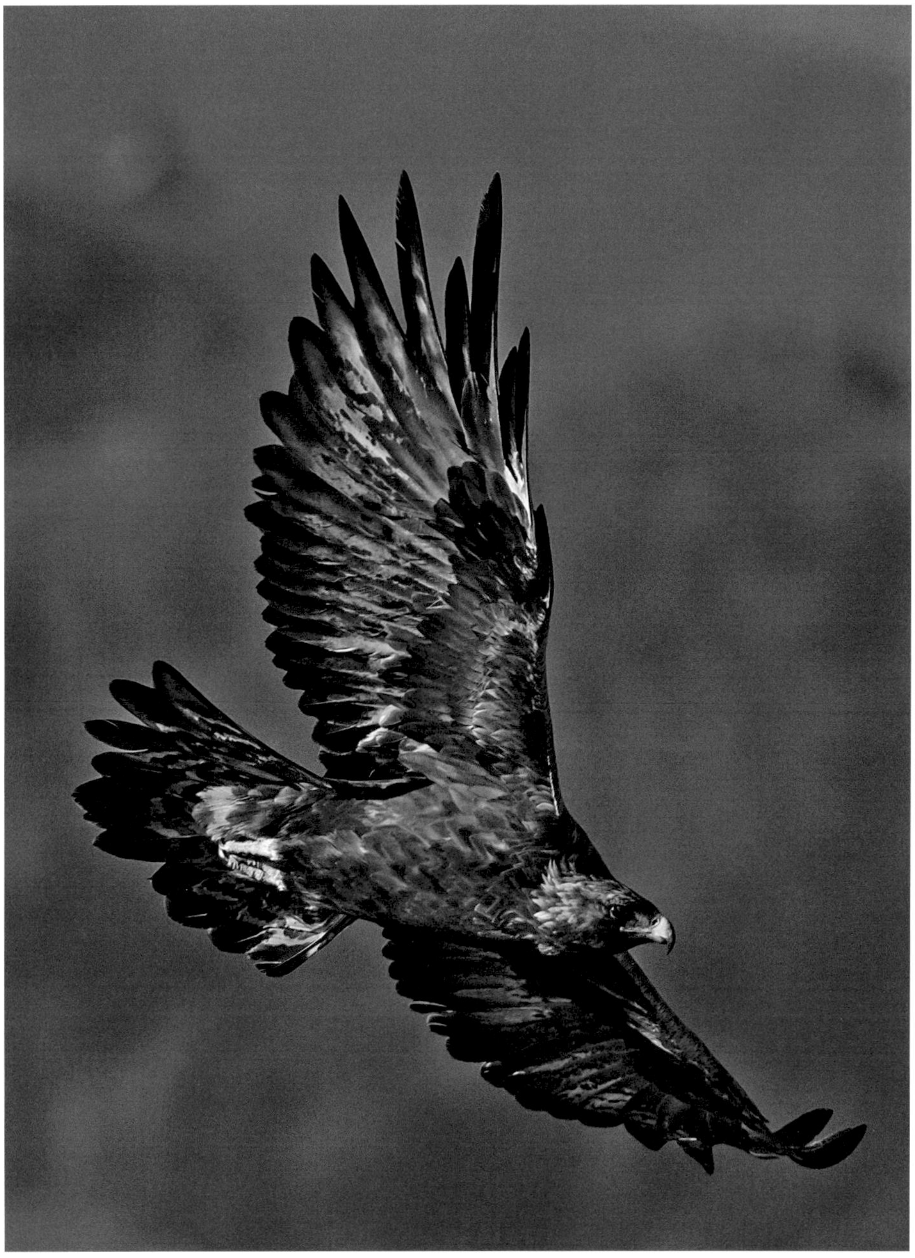

THE RED SKIRT IN THE SPYSPFAD

A FAIRY TALE FROM THE BERNESE OBERLAND IN SWITZERLAND. TRANSLATED AND SLIGHTLY MODIFIED \ 103 \

Long ago, golden eagles had their eyrie in the wild cliffs of the Sätze in the valley of Trümmleten. The voracious eagles soared over both passes of the Scheidegg as far as Kien and Lötschen in their raids for prey. Everything, whether fur or feather, including lambs and goatlings grazing on the meadows, were startled and ran for cover when the drawn-out "hiäh-hiäh" of the eagle call echoed through the valley. The mountain farmers from Lauterbrunnen never liked the bold bandits because of a gruesome incident linked to the eagles that has been passed down from generation to generation over centuries. A farmer from Mürren, from a family no one remembers today, was making hay with his wife at a place called Aegerti. It was a hot summer's day, and they put their two-year-old child, dressed in a red skirt, in the shade behind a small barn. Suddenly, a golden eagle stooped down from high-up in the sky, snatched the small child with its talons, and flew with her over the cliff of the Mürrenfluh, and onwards into the void.

Everything happened so quickly that the two farmers only saw the feathered predator after it had already soared high over the valley with their child in its claws. They watched the eagle place the innocent child on a small path in the midst of the Stellifluh, which was several hundred meters high, right under the reddish-colored part of the wall. The cliff ledge was completely unreachable for any human being and neither the farmers from Aegerten nor any inhabitant from Mürren could help. So exposed was the small path in the large rock wall that even God would have been powerless to help! With horror, the farmer couple watched how the golden eagle sat close to the child and devoured its prey.

Even many years after the sad incident, when the snow on the ledge had melted, one could spot the remains of the small red skirt from the poor child, glowing in the evening sun. Since this unfortunate occurrence, the ledge with the path in the middle of the dizzying, vertical cliff of the Schwarzmönch has been called the "Spyspfad" in the local Swiss dialect, meaning the Food Path.

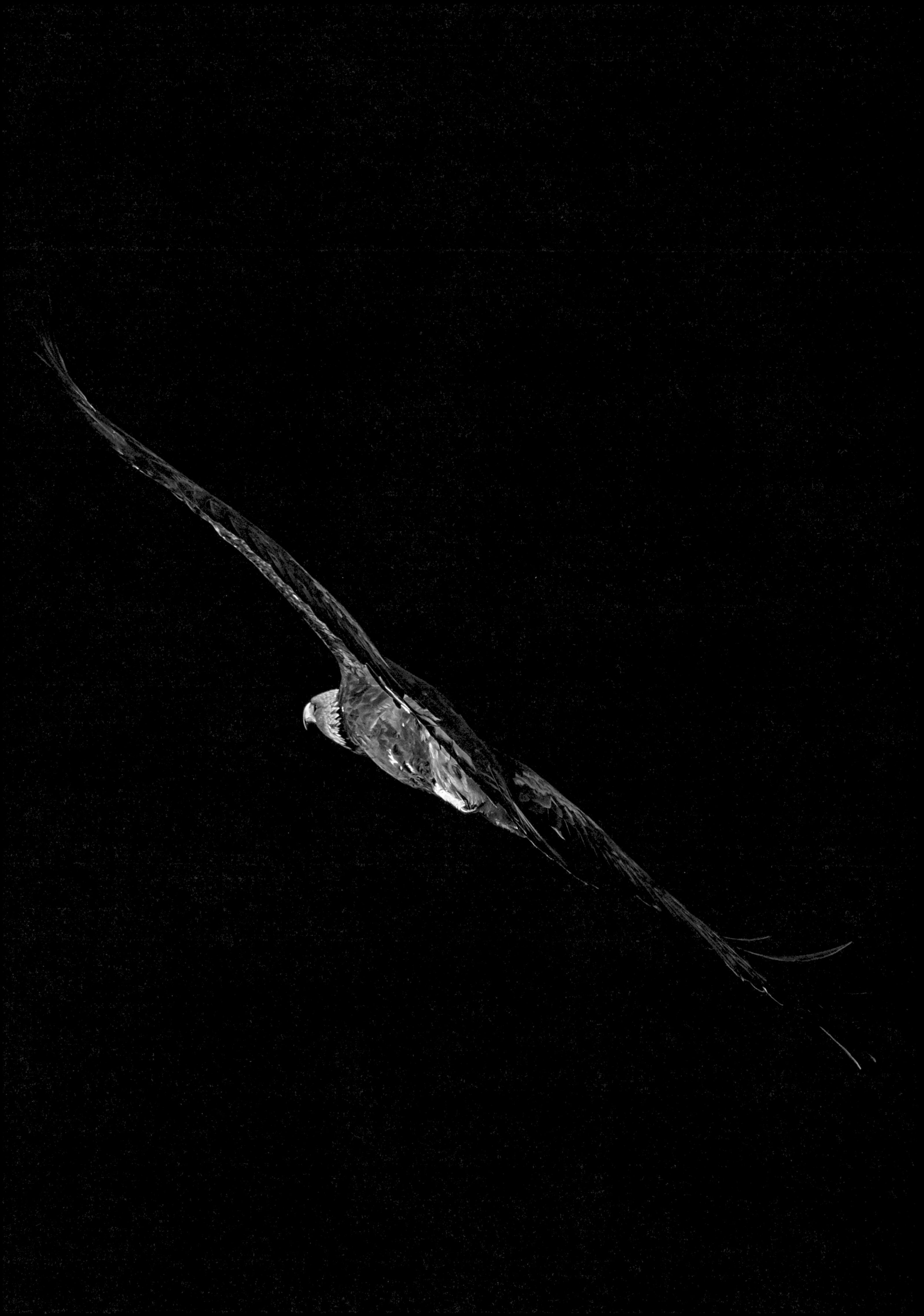

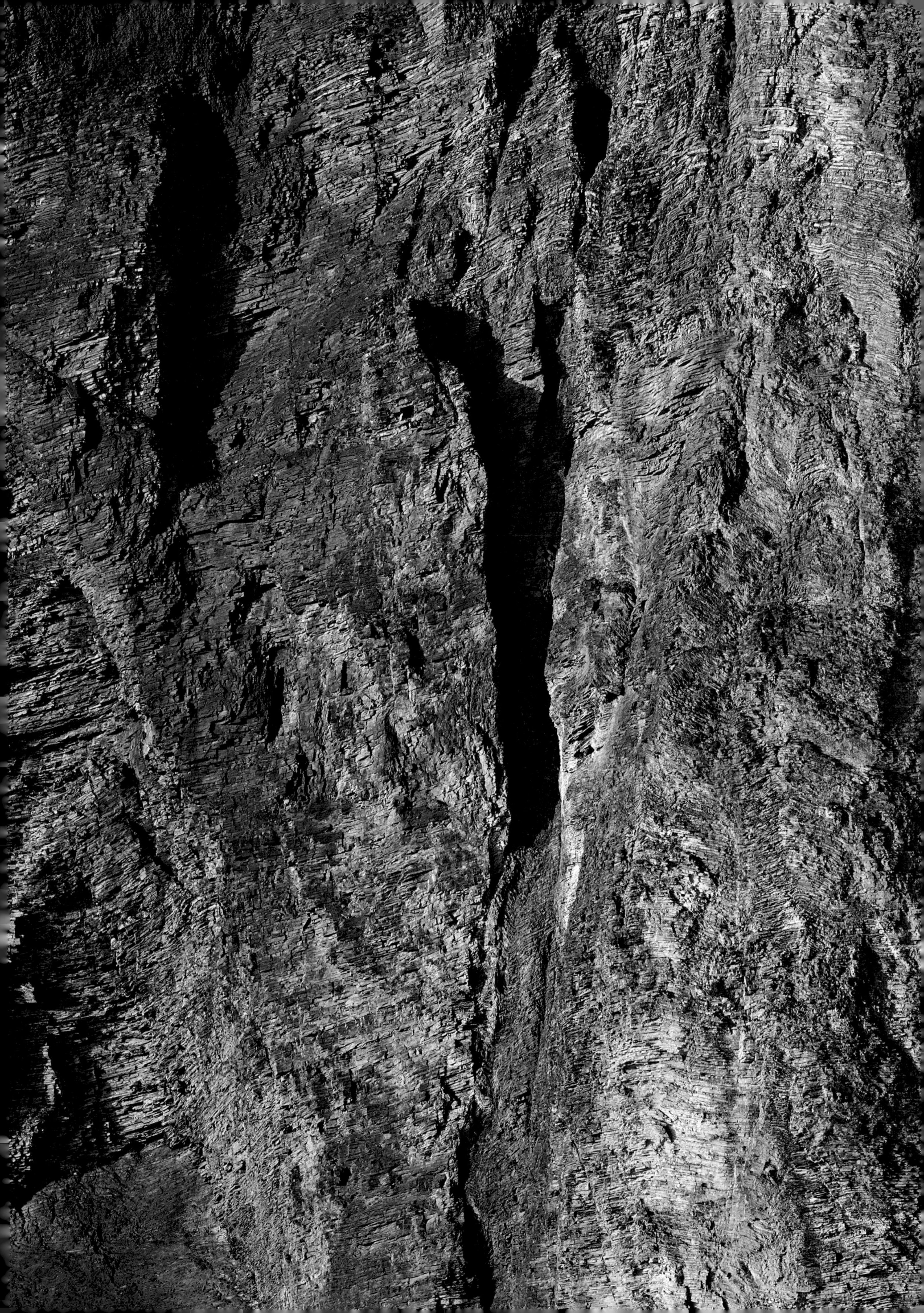

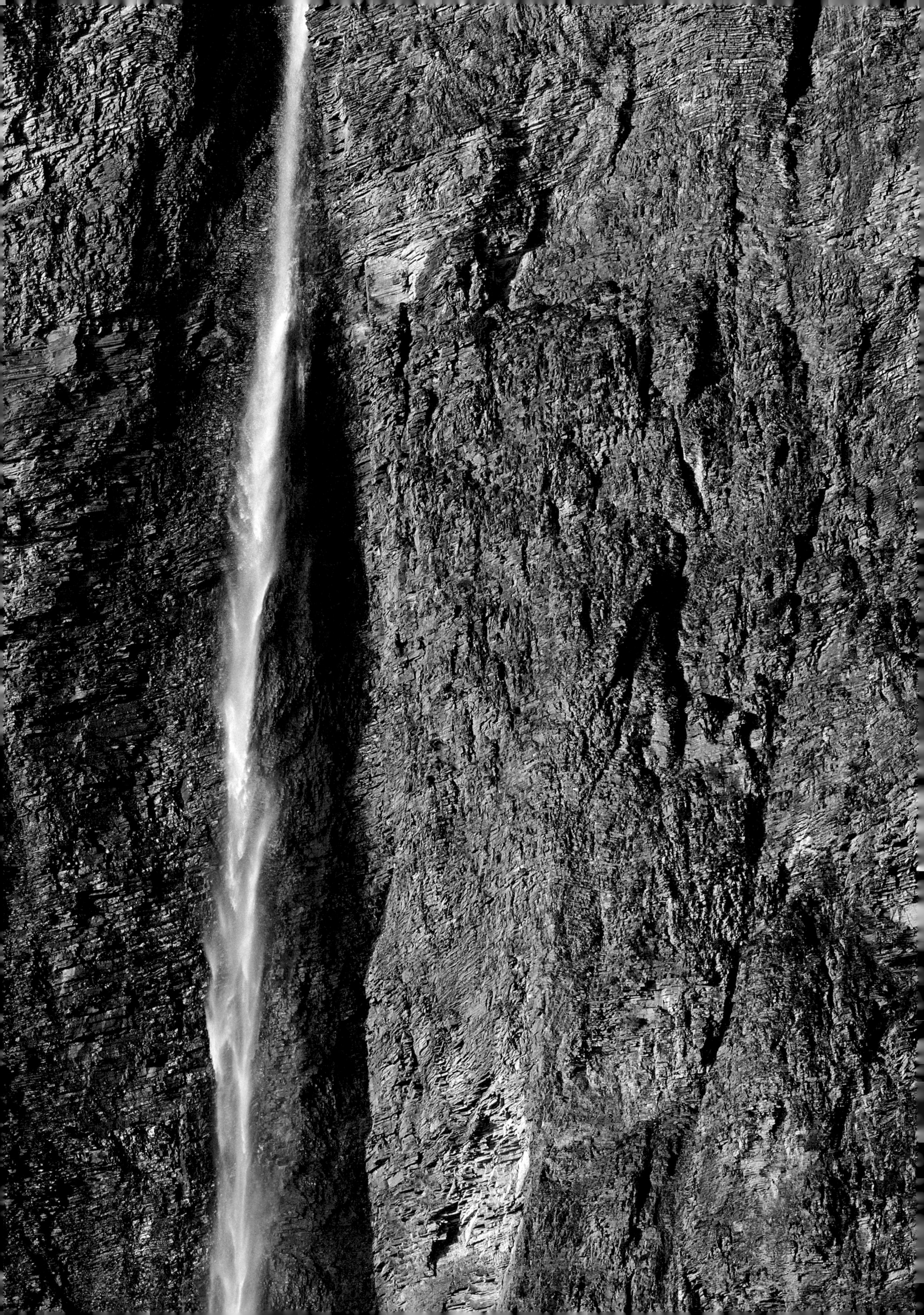

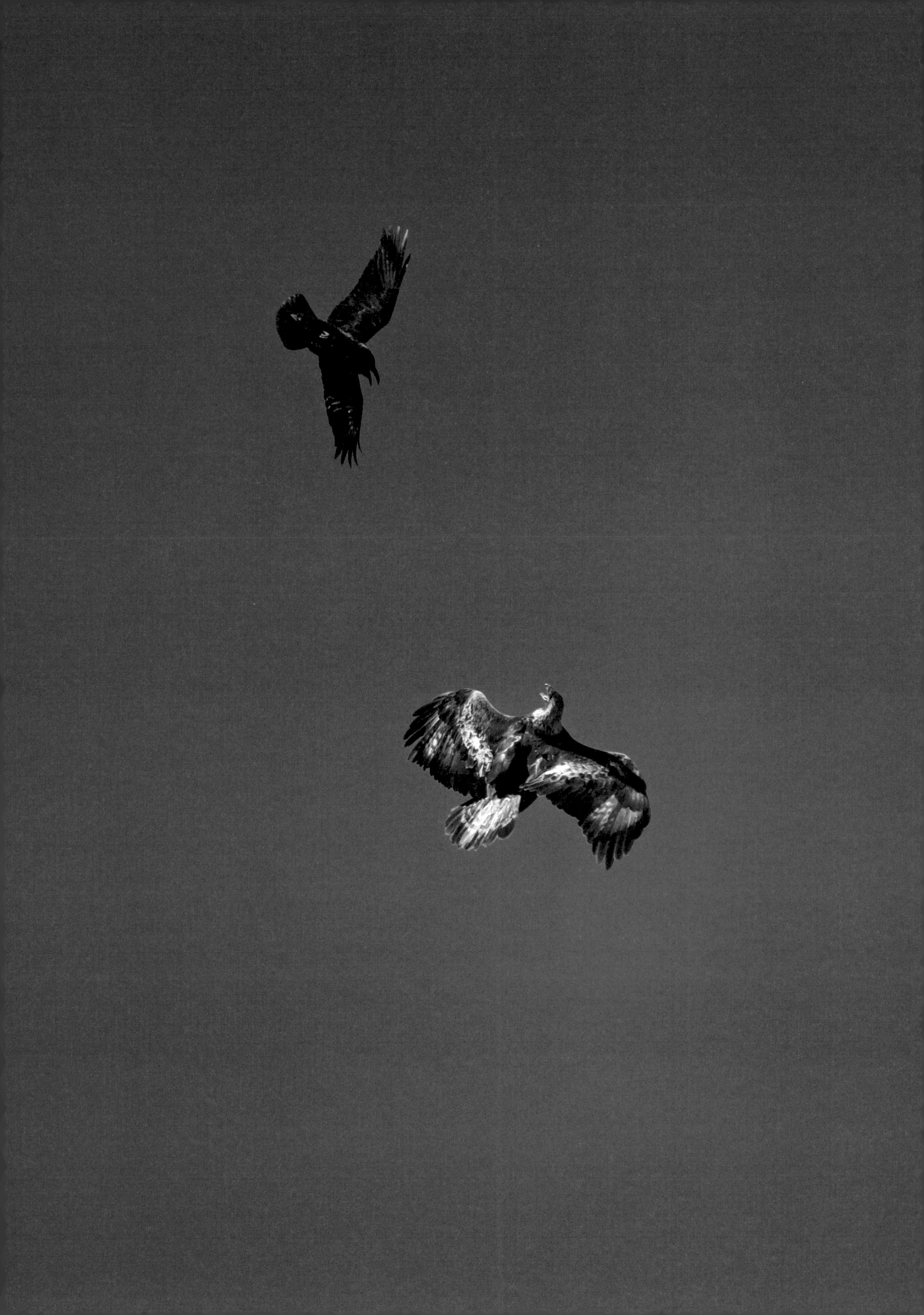

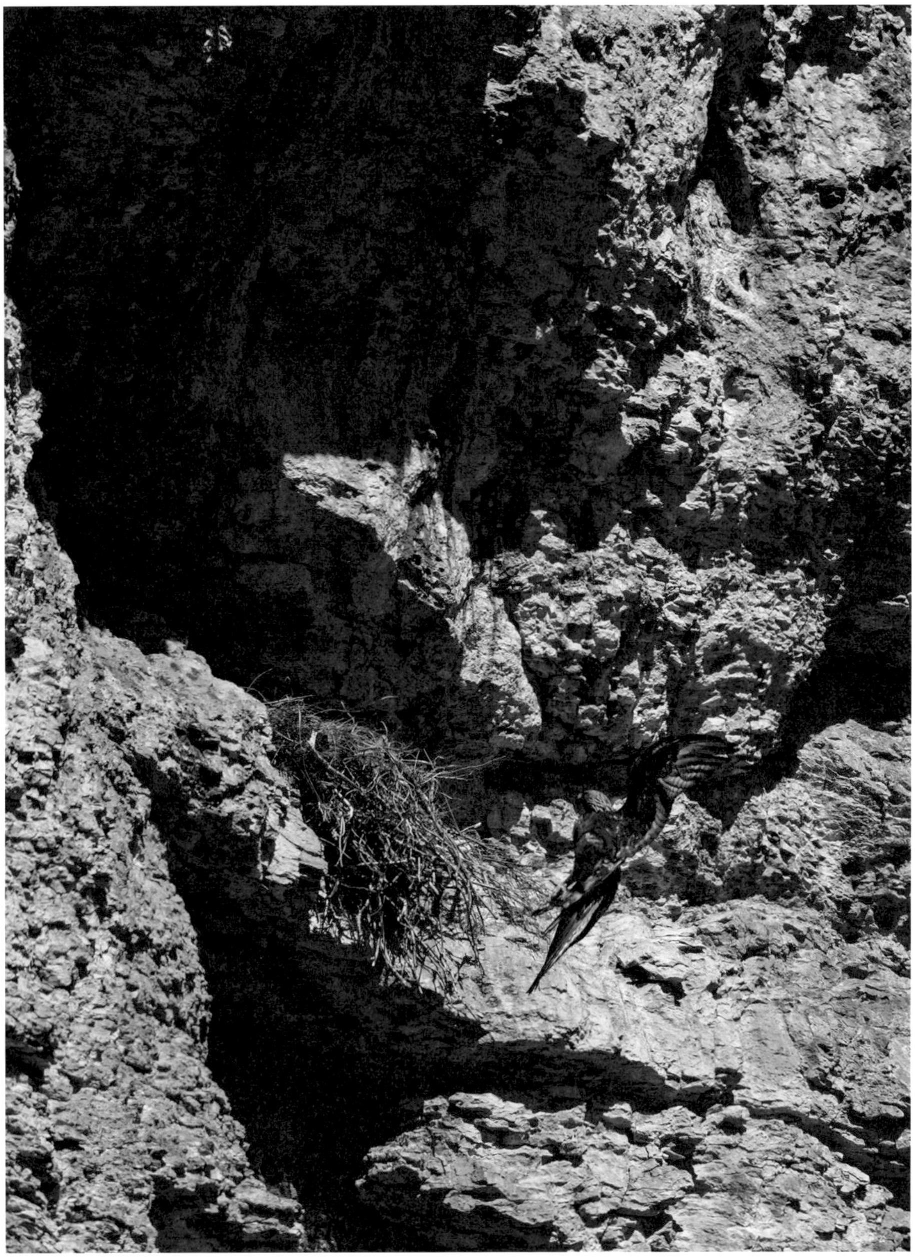

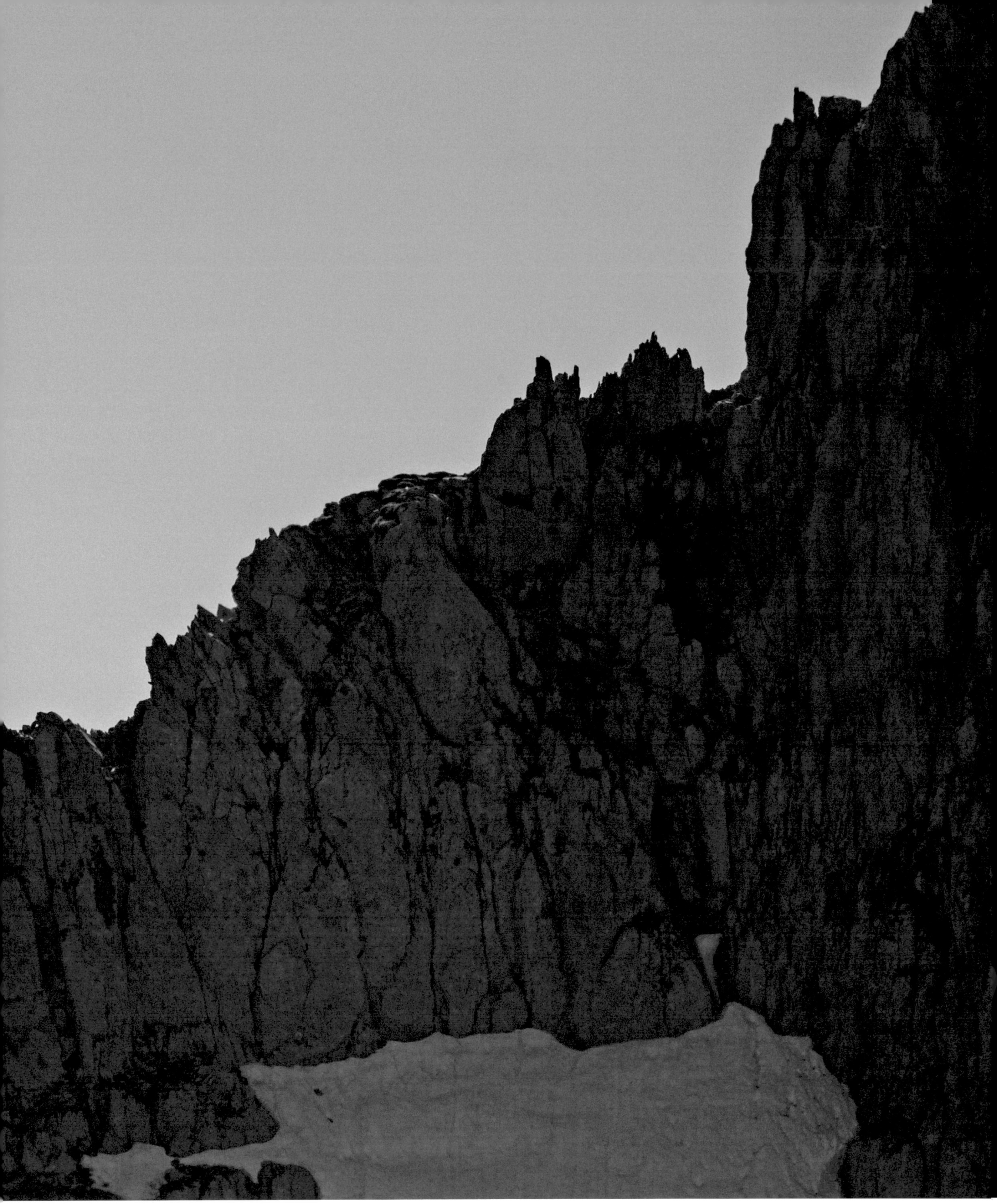

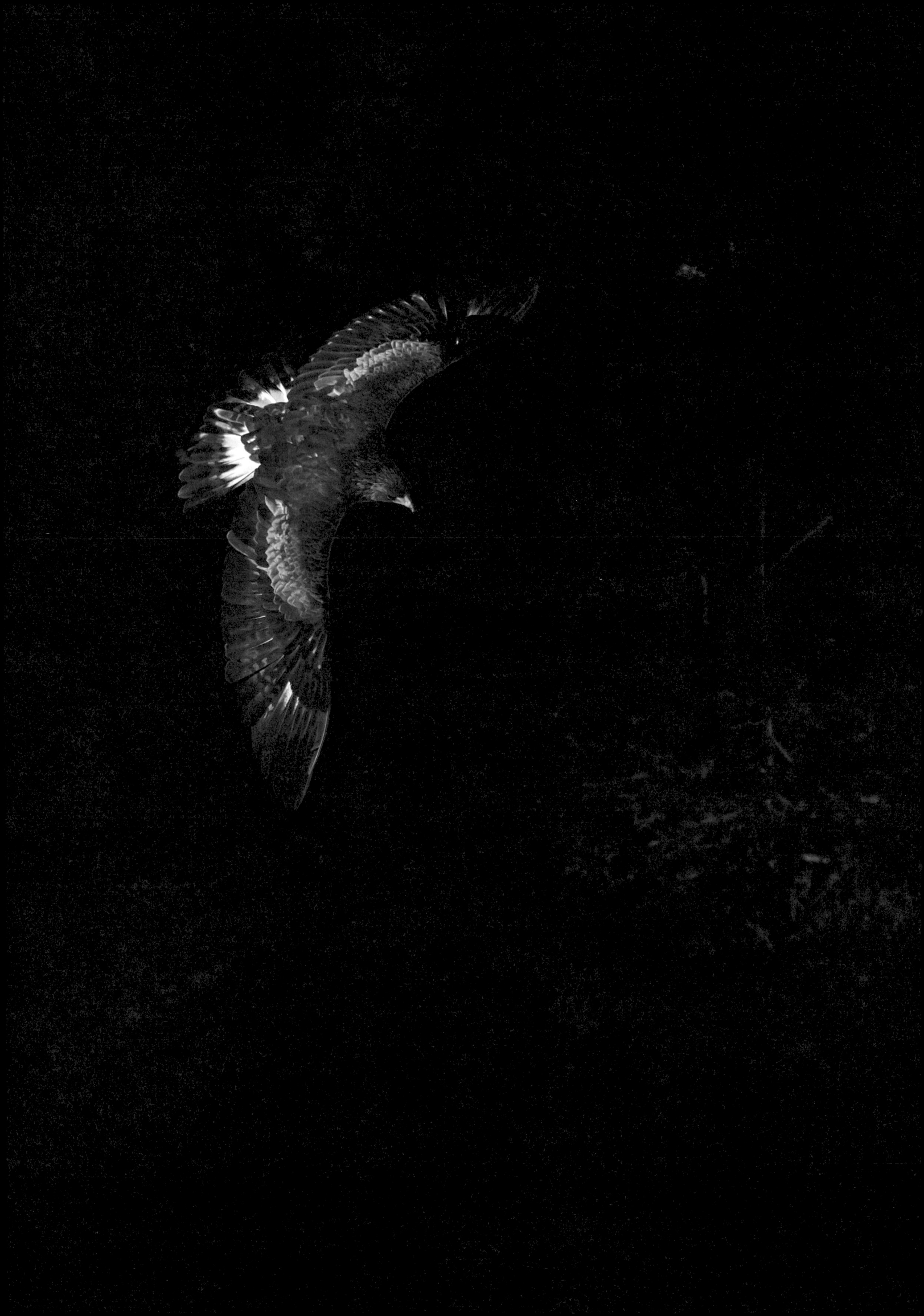

HOW THE EAGLE DOES

TIMOTHY OTIS PAINE \
1824 — 1895 \ 104 \

The Eagle

How the eagle does: —
Gathering up his might,
Quitting where he was,
Soars he in the height.
But his aerie home
Is not always grand:
Now on mountain dome,
Now in lowly land.
In a rugged wold,
Be it but apart,
He shall build his hold,
Take his mighty start.
Where he makes his bed,
Where he piles his lair,
Turns his noble head,
'T is the king that's there.

Where he heaps his nest,
Where he lies in state,
Where he takes his rest,
There the place is great.
When he looketh far
Through the forest dim
From a naked spar,
Then look up at him.
Feel him seize thine eye;
See him once for aye;
Watch him towering high
On his spiral way,
Till, a little mote,
Black upon the blue,
He is like a boat
Sailing out of view.

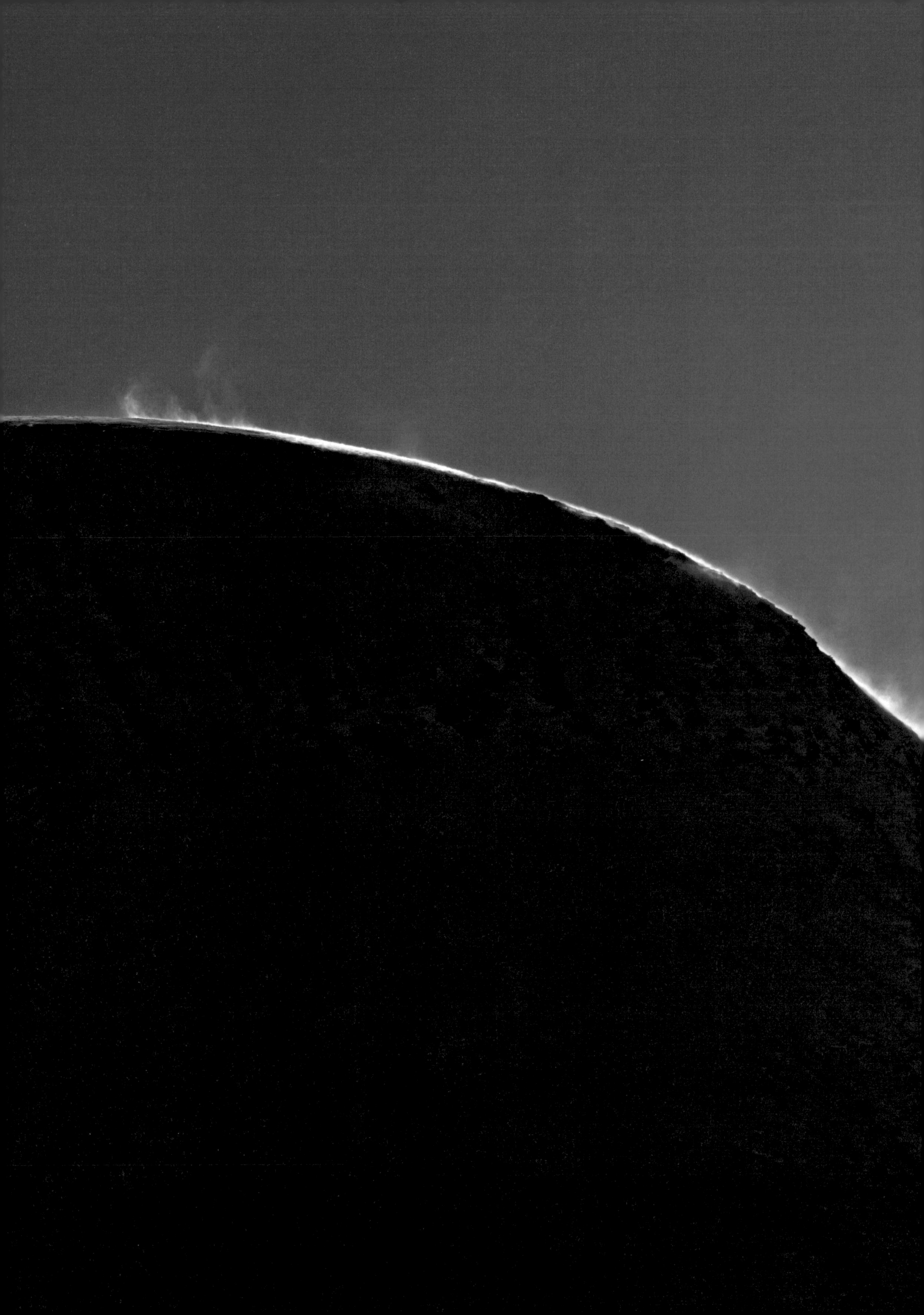

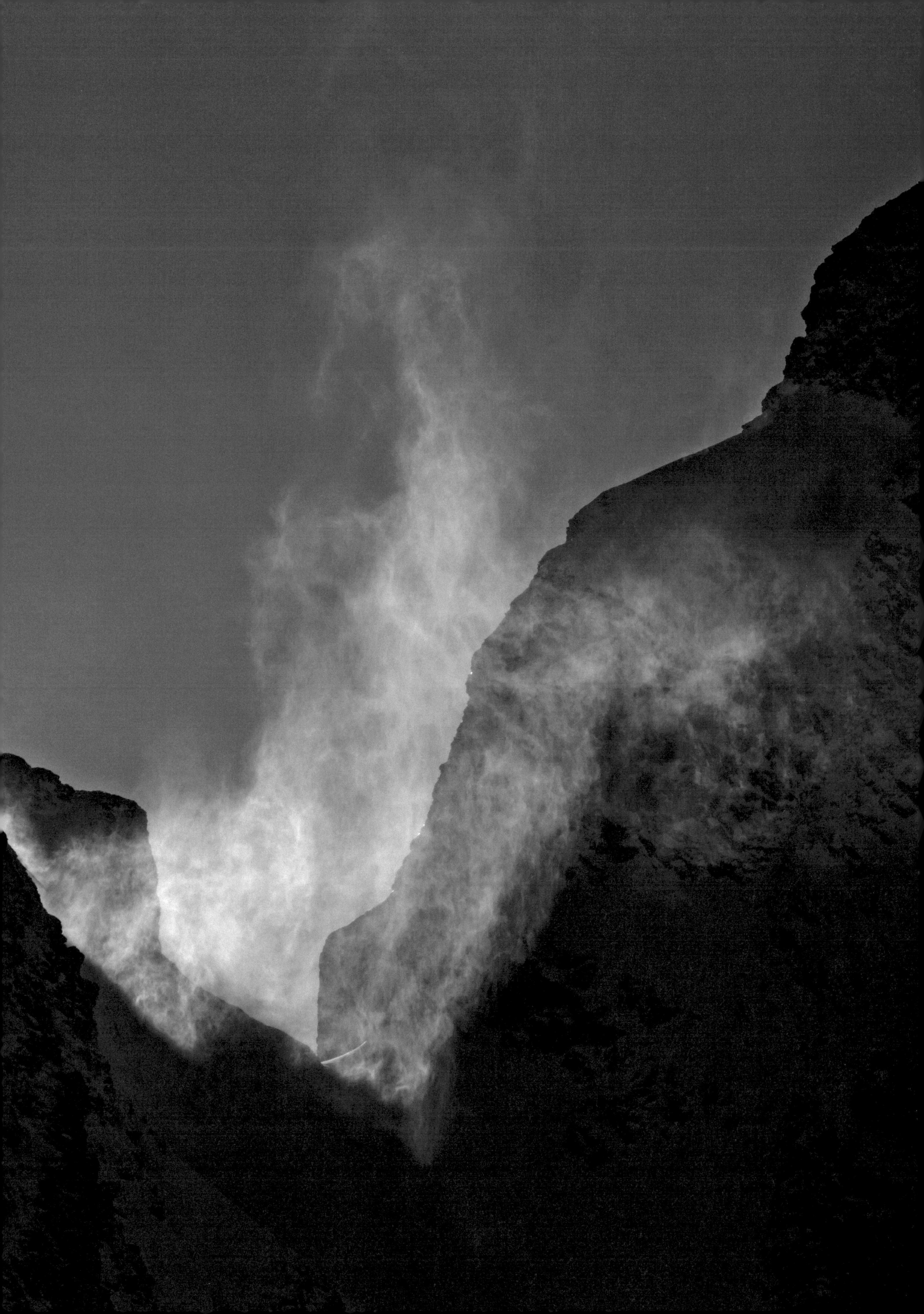

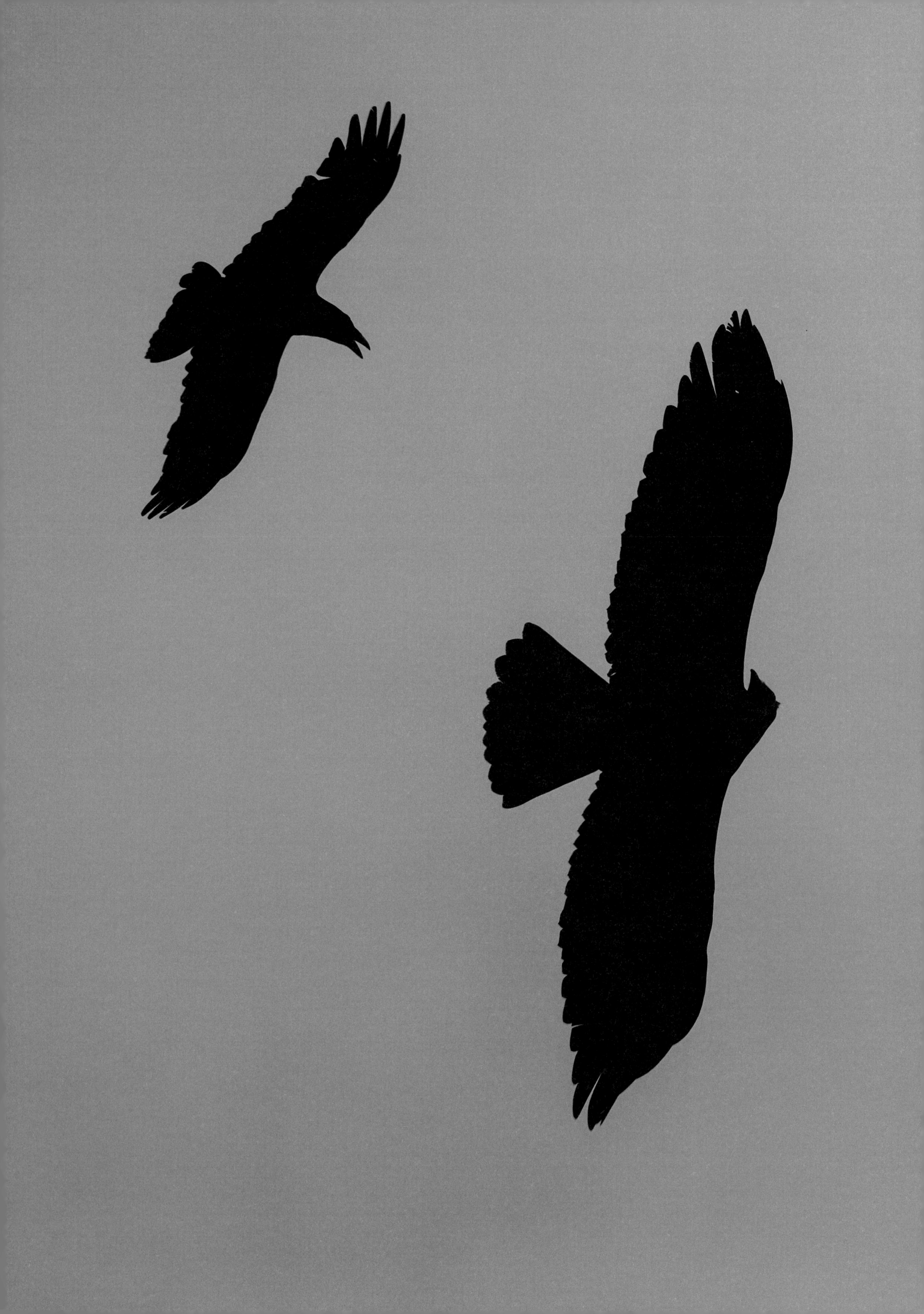

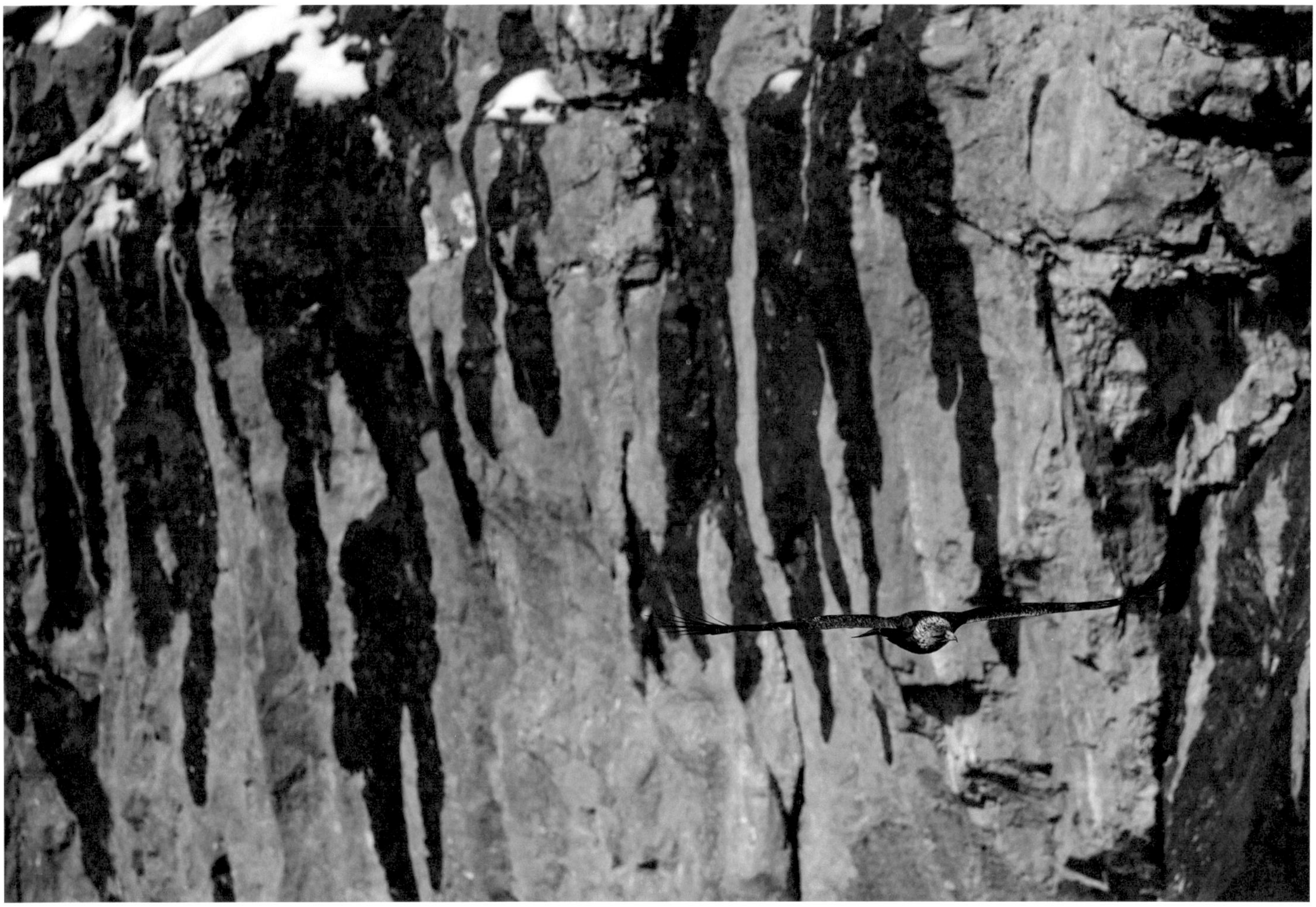

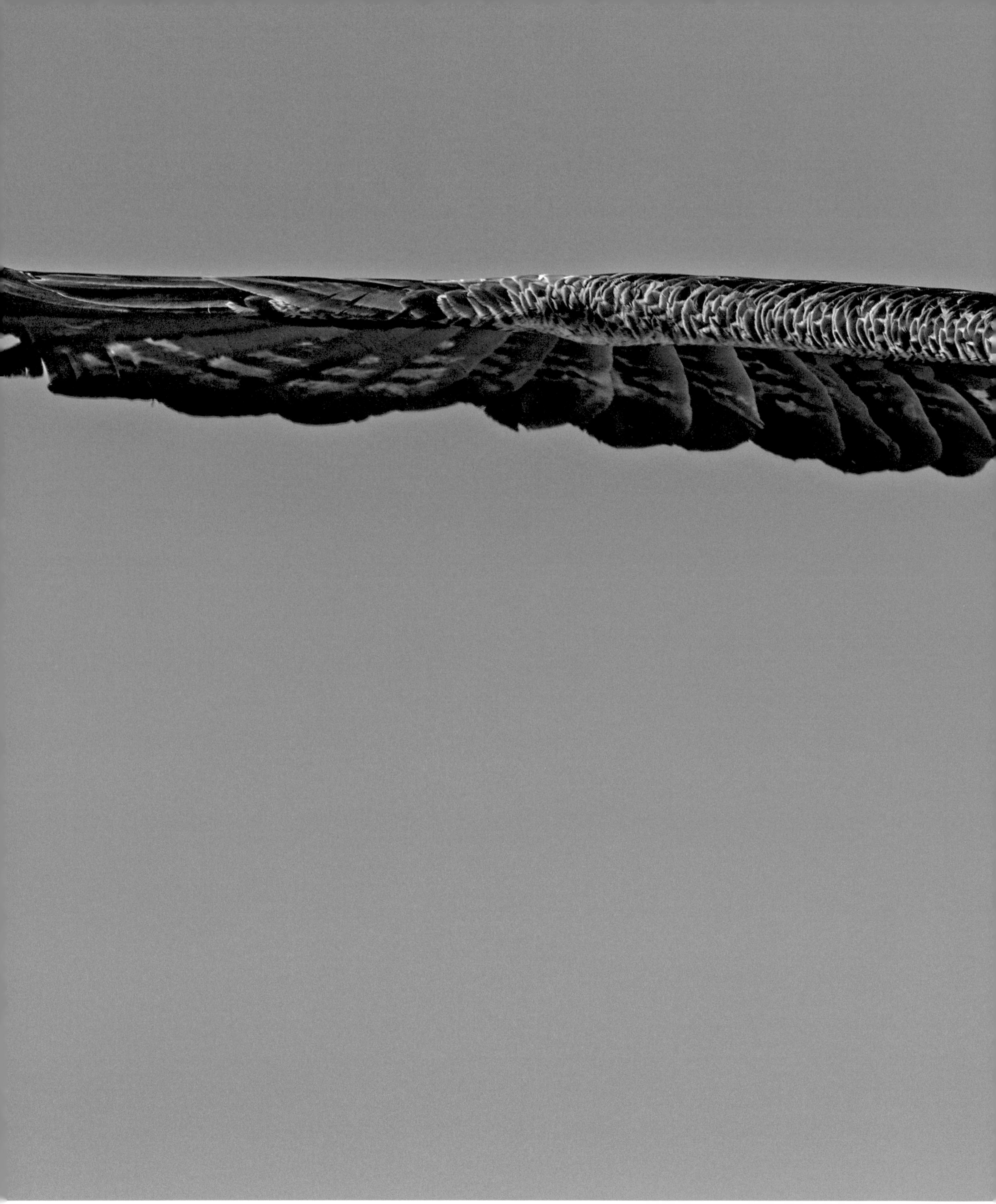

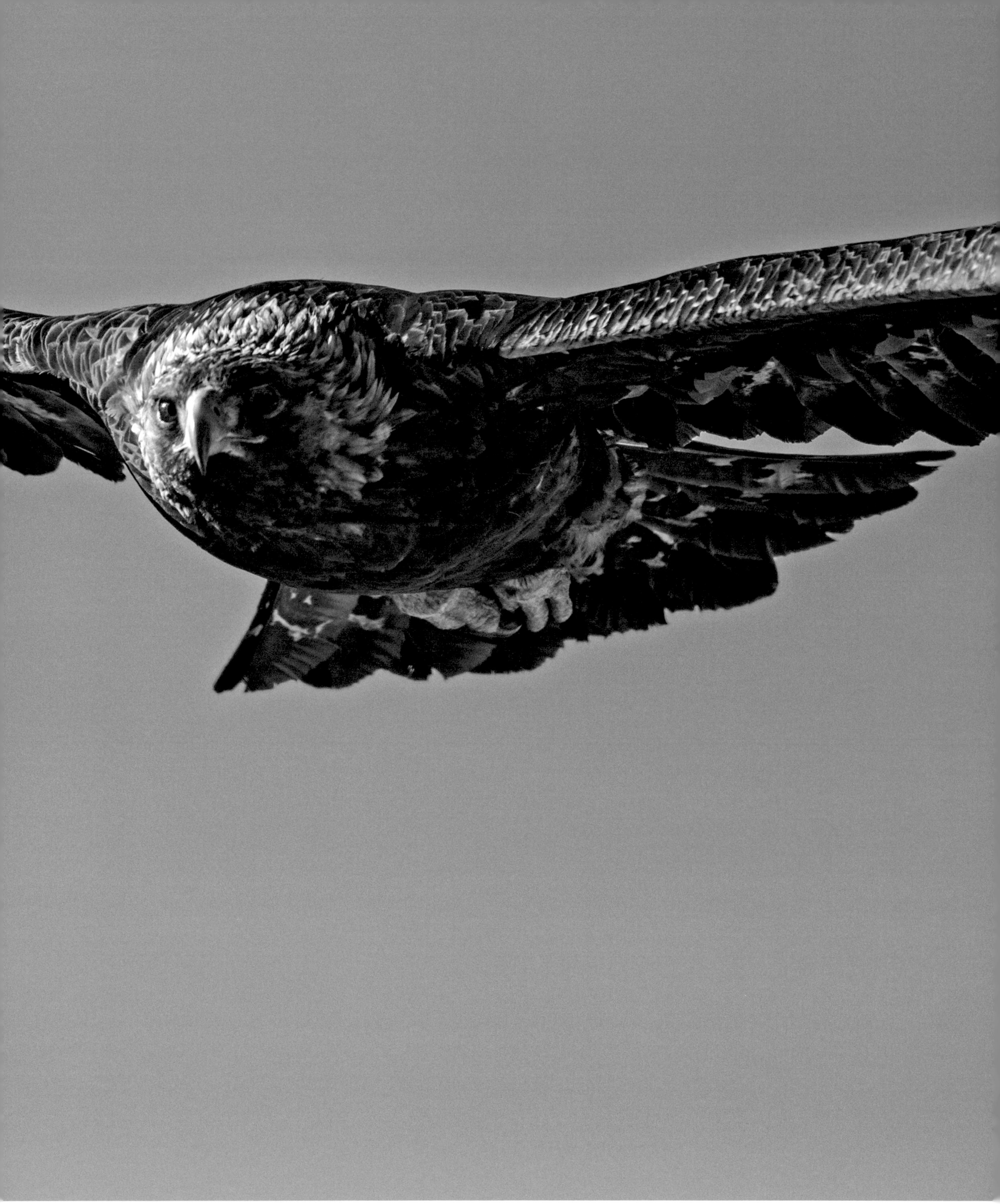

THE EAGLE BRIDE

A FAIRY TALE FROM THE GRISONS IN SWITZERLAND. TRANSLATED AND SLIGHTLY MODIFIED \ 105 \

A poor, married couple once lived in a large city and struggled for a long time to have children. When god finally gave them a small boy, the clever midwife suggested choosing the first horseman that would come along the path as the child's godfather. The boy's father followed her advice and stepped onto the road where he waited for the first horseman. When a proud man on his horse came along the road, he asked him to become the godfather of his son. The stranger agreed, lifted the small boy out of the crib, and presented himself as the king of a large empire of several islands. The wealthy nobleman liked the little boy so much that he gave the parents a great amount of gold, with the order to take great care of the child. He told them to send the boy to his court once he had reached the age of eighteen years, so that he could marry his own daughter.

The parents fulfilled the king's wish and once the boy had celebrated his eighteenth birthday, they sent him to his royal-blooded godfather, king of the island empire. On his way, close to a spring, the young man met an ugly looking dwarf, who threatened to kill him if he would not do what he was told. The godchild of the king had no choice and promised to present the dwarf as the future groom of the princess, while he himself would be introduced as his servant. When the two arrived at the king's empire the powerful ruler was very surprised to find his supposedly lovely godchild turned into a midget with a wobbly, miss-shaped head and skinny, crooked legs. Also, the princess did not like the bandy-legged groom and rather admired the tall and beautiful servant. Hence, the dwarf became so jealous that he cursed and banished the king's daughter to the island of darkness in the vast ocean, where the sun never shone. In addition, he was of such evil spirit that he presented the servant as a sorcerer, so that the king sent him to prison where he should be sentenced to death by the wheel.

But in the night before the execution of the poor boy, an honorable old man appeared and advised him to ask, as his last wish, for three meat-loaded ships to enter the sea and search for the king's daughter. The youth followed this advice, stepped in front of the king, and vowed not to rest until the princess was found, should his wish be fulfilled. Hopeful to get his daughter back, the mighty king agreed to his plea.

The youth embarked on the ship and sailed out into the rough sea, where he first landed on the island of the bears, whom he gave a shipload of meat, as they requested. He asked the king of the bears about the island of darkness, but the wise old bear did not know where it lay. Nevertheless, he assured the youth that upon a single whistle, the bears would come to help if the young man was in trouble. He then sailed to the island of leopards, where he had the very same experience as on his first stop. The leopards demanded the second load of meat but did not know where the island of darkness was located. Similarly, the leopards promised immediate help in case the young man was in need. On the island of the eagles, the youth had more luck.

The king of the eagles was delighted with the shipload of meat, and so he granted the young man the gift to change into an eagle whenever he wished. An old eagle that was also present told the godchild of the king that he had once visited the island of darkness, and that he would be happy to accompany the youth on his voyage. Delighted, the young fellow, who was on a mission to find the king's daughter, and thus save his own life, agreed.

Hence, he set sail with his new companion out into the ocean, where they landed on the island of darkness. They first encountered an old woman with seven white mice. The old eagle asked her where the castle with the imprisoned and spellbound princess might be found. She readily explained where it was and insisted that they take the seven mice along to guide them to the castle. But the old eagle, used to the bright sun, and having had enough of the darkness, opened his large wings and flew south towards his bright home. As promised, the small mice led the youth to the castle that lay at the tip of a very steep rock. The tall castle had just one single window high up under the roof, from where a pale and noble visage looked out into the everlasting darkness. At this moment, the godchild of the king turned himself into an eagle, and flew up high to enter the window of the small chamber, where he regained his original form to the great delight of the beautiful imprisoned princess. But the king's daughter told the brave young man that before they could leave and be free, he first must kill the fearsome dragon that guarded her, and who lived in the deepest, darkest part of the castle. The young fellow did not hesitate, took the sword into his hand, and descended the many stairs that led down into the dungeons where the monster lived. Once he had discovered the terrifying dragon, he swung his sword and struck the creature with a powerful blow. But unfortunately, on striking the scaly armor of the monster, the blade shattered into pieces as if it were a light stick. At the mercy of the dragon, and close to death, the youth suddenly remembered the promises of his new friends and whistled to call them for help. The echo had hardly trailed away, when from all sides bears, leopards, and eagles appeared to do battle with the terrifying monster who, after a long fight, succumbed to the wounds inflicted by the united army during the hefty battle. Rejoicing, they brought the king's daughter to the ship and burned the remains of the powerful dragon, whose ash they threw into the sea. At this very moment the many bears, leopards, and eagles that had saved his life, suddenly changed into human figures, and formed a splendid entourage of friends around the brave, young man.

The sun rose over the horizon, shedding a warm light onto the island of darkness, while the castle fell into ruins. The king's godchild and the princess, together with the liberated knights and noblemen, sailed south towards the large empire where they were received with jubilation. The treacherous dwarf was sentenced to death by the wheel, on command of the mighty king.

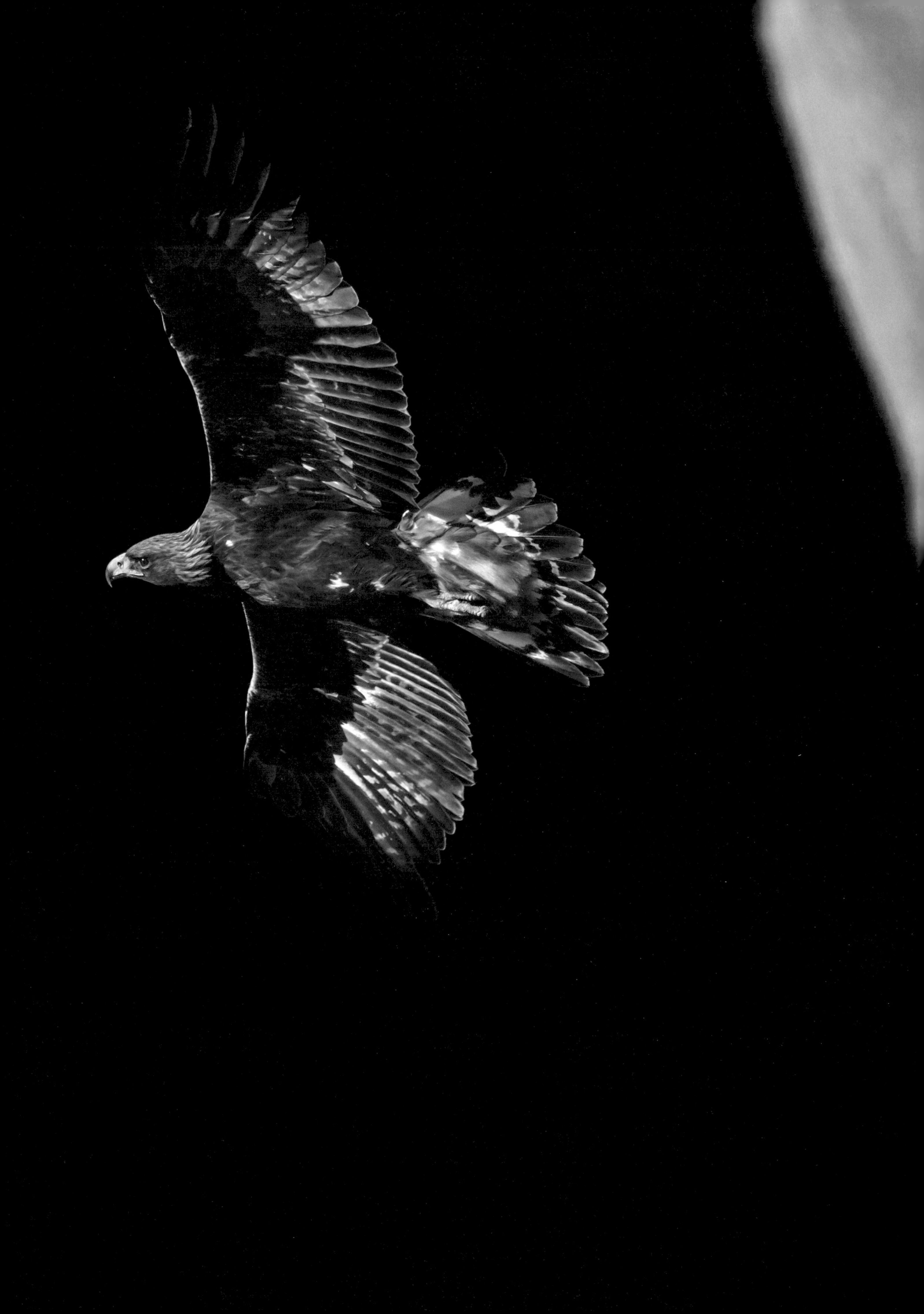

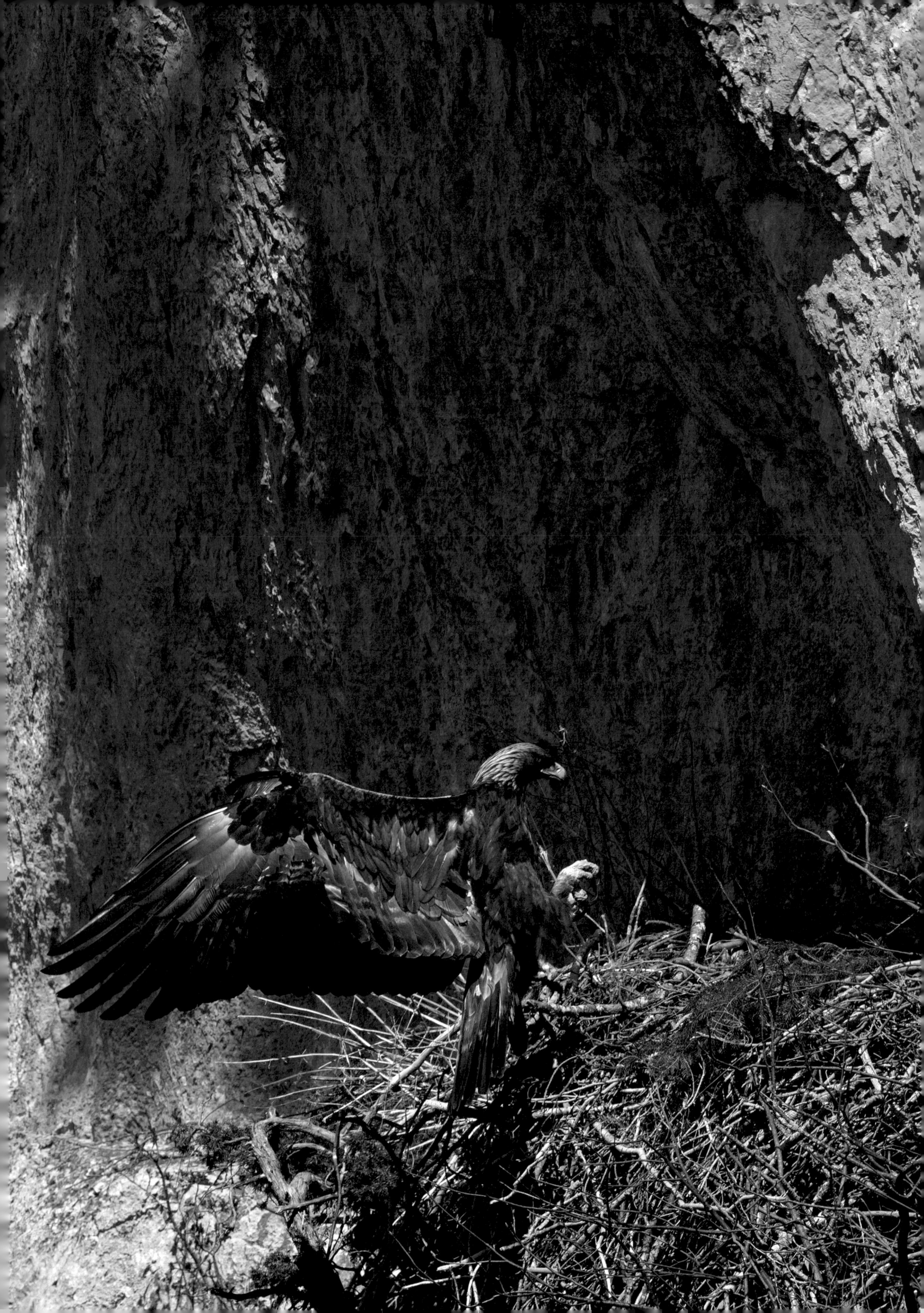

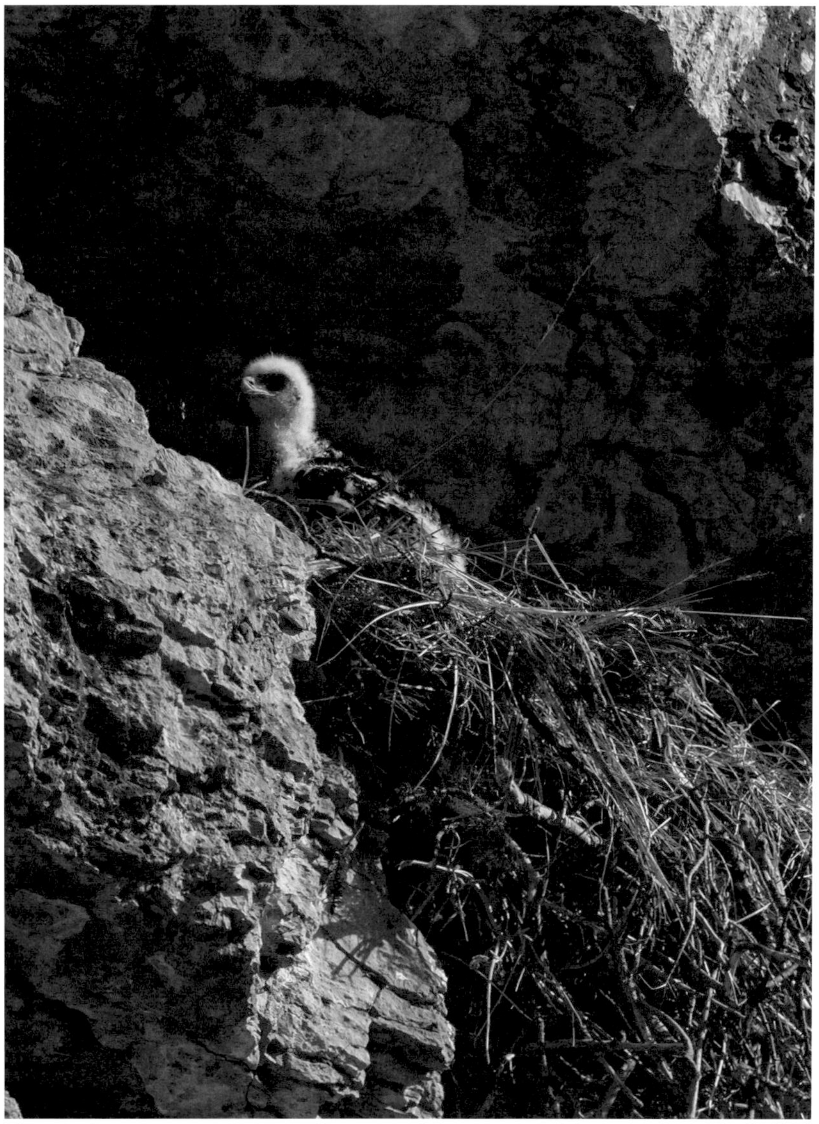

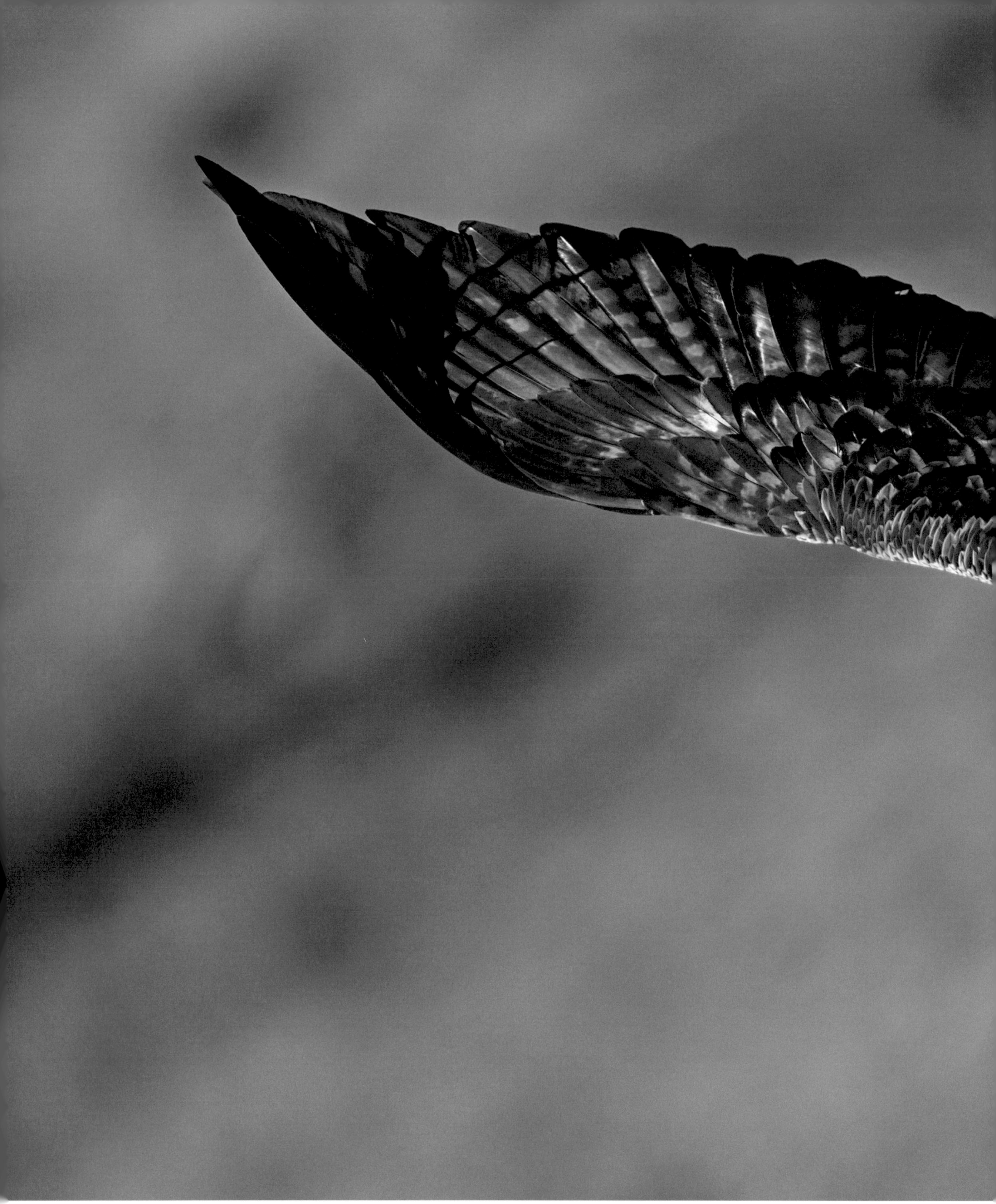

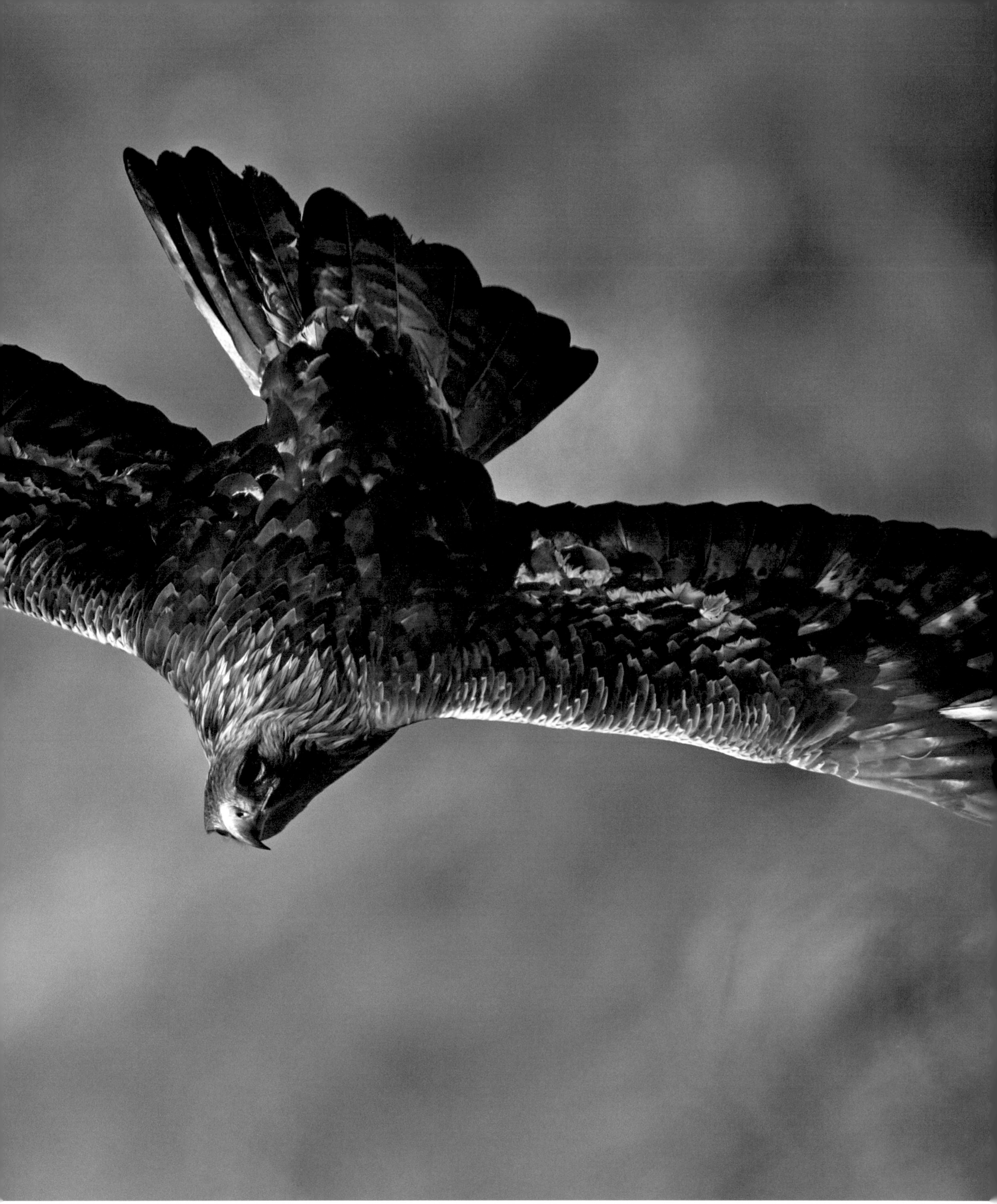

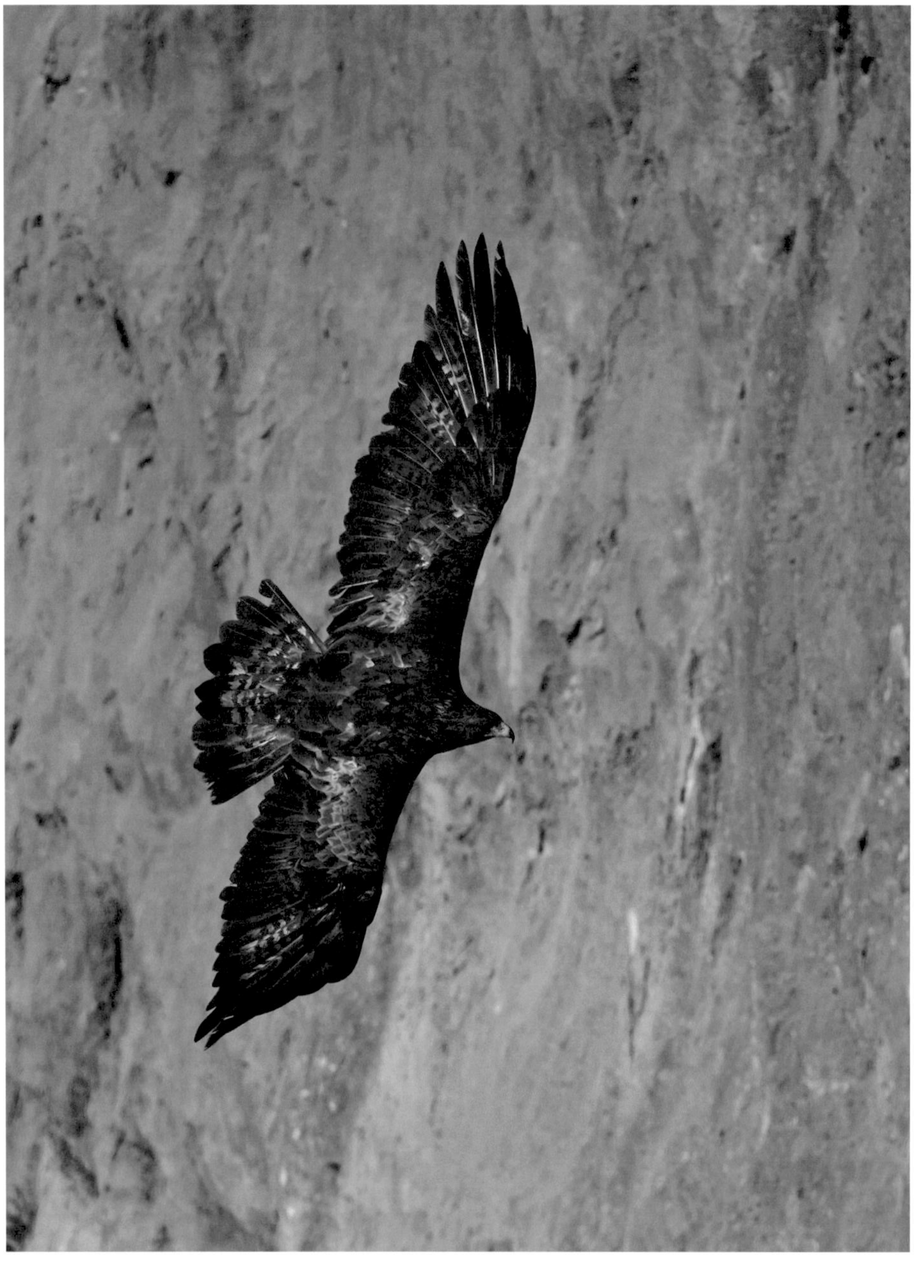

LIKE A FRAGMENT OF A CLOUD

ALFRED BILLINGS STREET \
1811 — 1881 \ 106 \

The Eagle

We touch the green marge; hark! a shriek shrill and loud,
    A bird with huge wings, like a fragment of cloud,
Shoots swift from the gorge, sweeps around, then on high
    Cleaves his way, till he seems a dim spot in the sky;
Then stooping in circles, contracting his rings,
    He swoops to a pine-top and settles his wings;
An eagle! an eagle! how kingly his form!
    He seems fit to revel in sunshine and storm;
What terrible talons, what strength in that beak,
    His red, rolling eye-balls the proud monarch speak;
He casts looks, superb and majestical, down,
    His pine for a throne, and his crest for a crown;
He stirs not a feather, though shoutings arise,
    But still flings beneath mute contempt at our cries;
A branch is hurl'd upward, whirls near him, but vain,
    He looks down his eloquent, glorious disdain,
Till he chooses to spread his broad pinions of gray
    And launch in majestic, slow motion away.

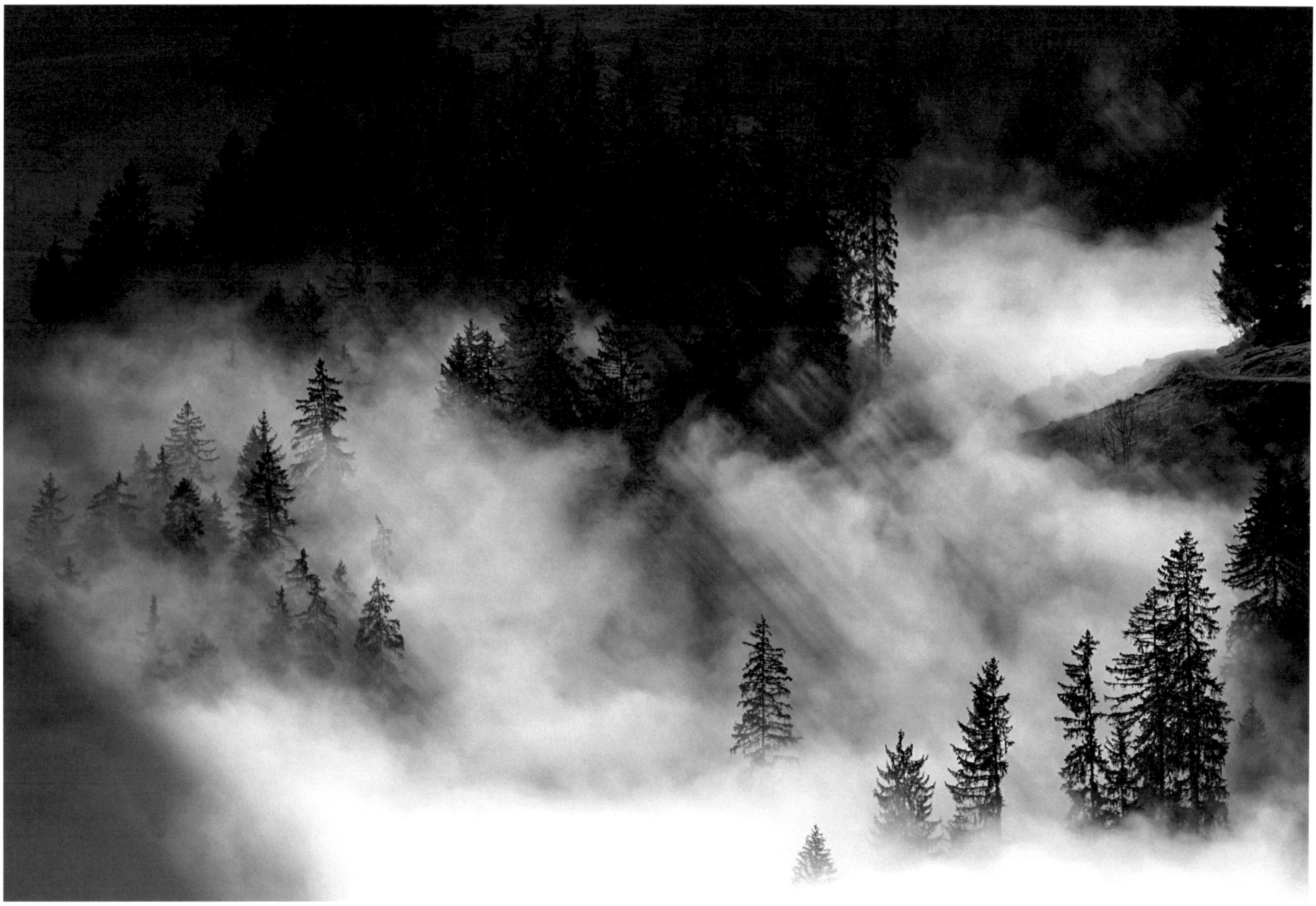

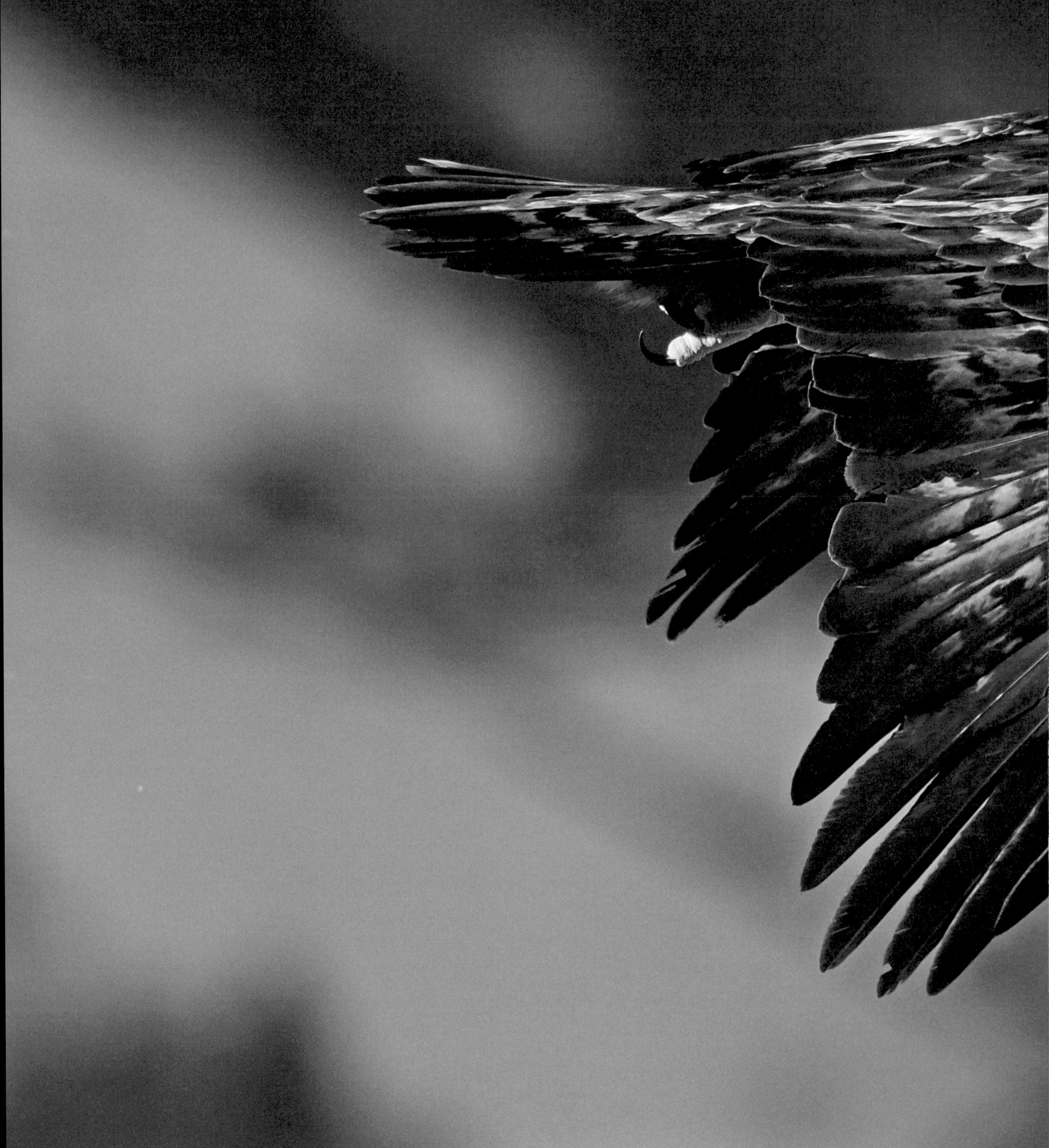

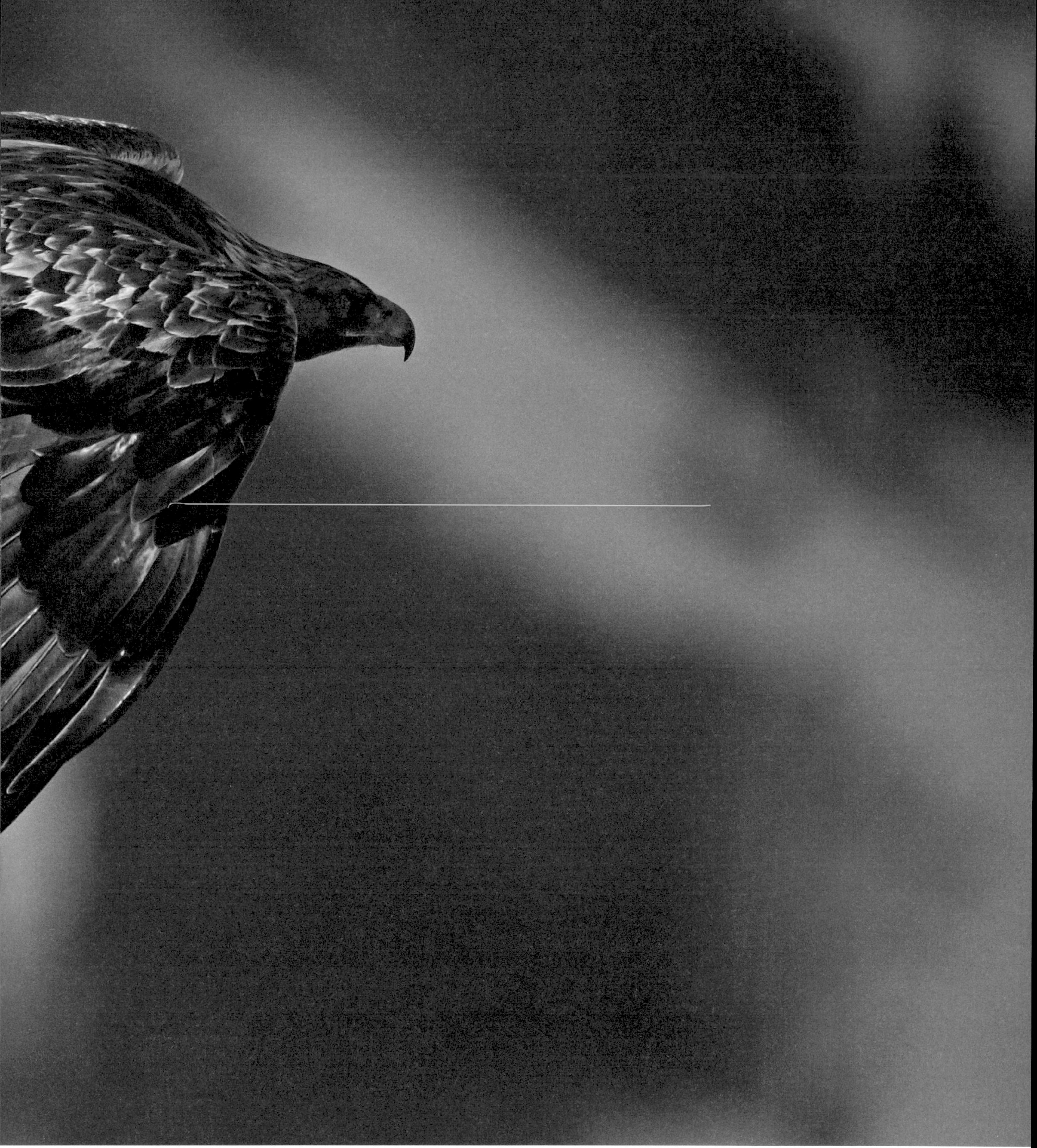

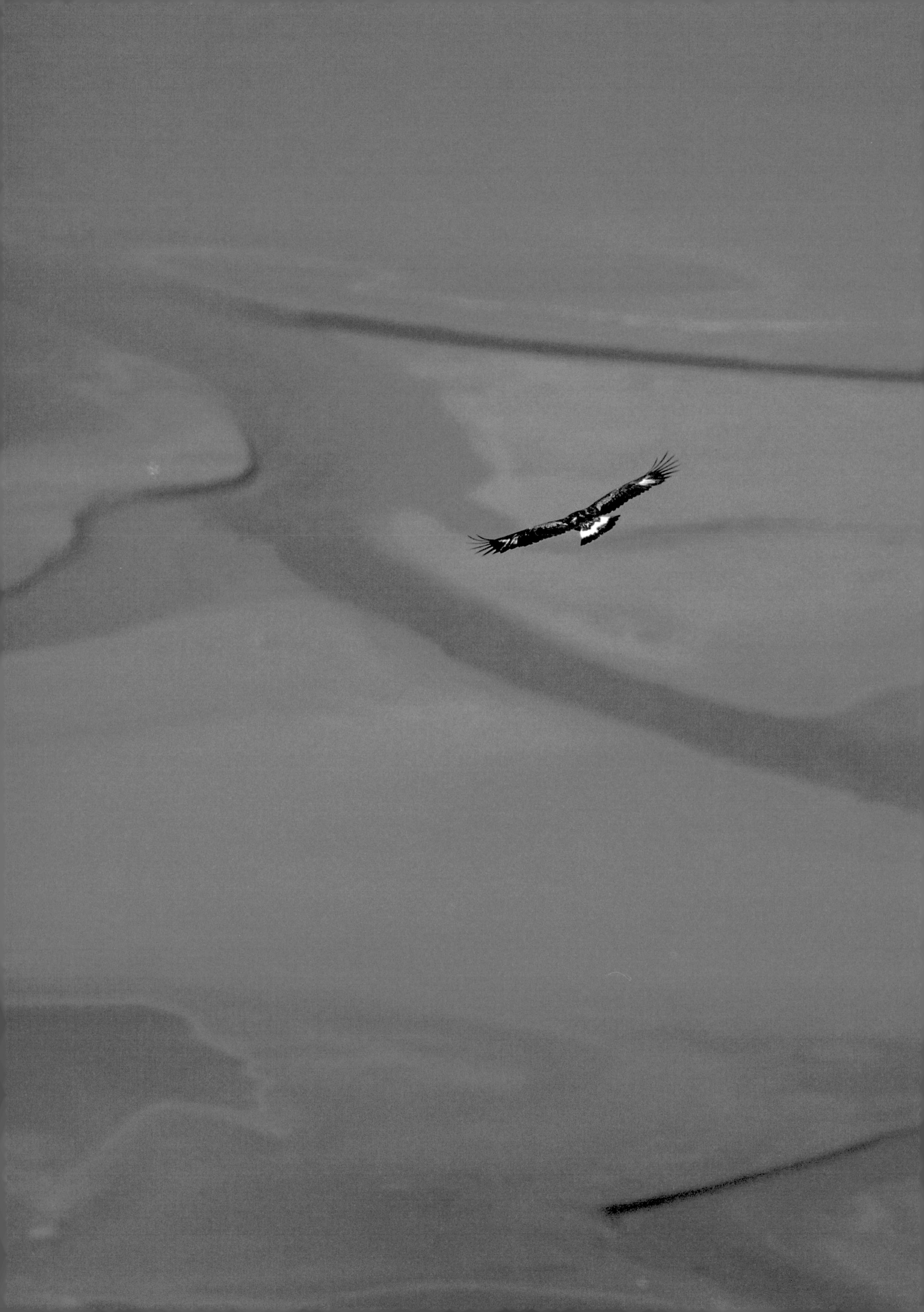

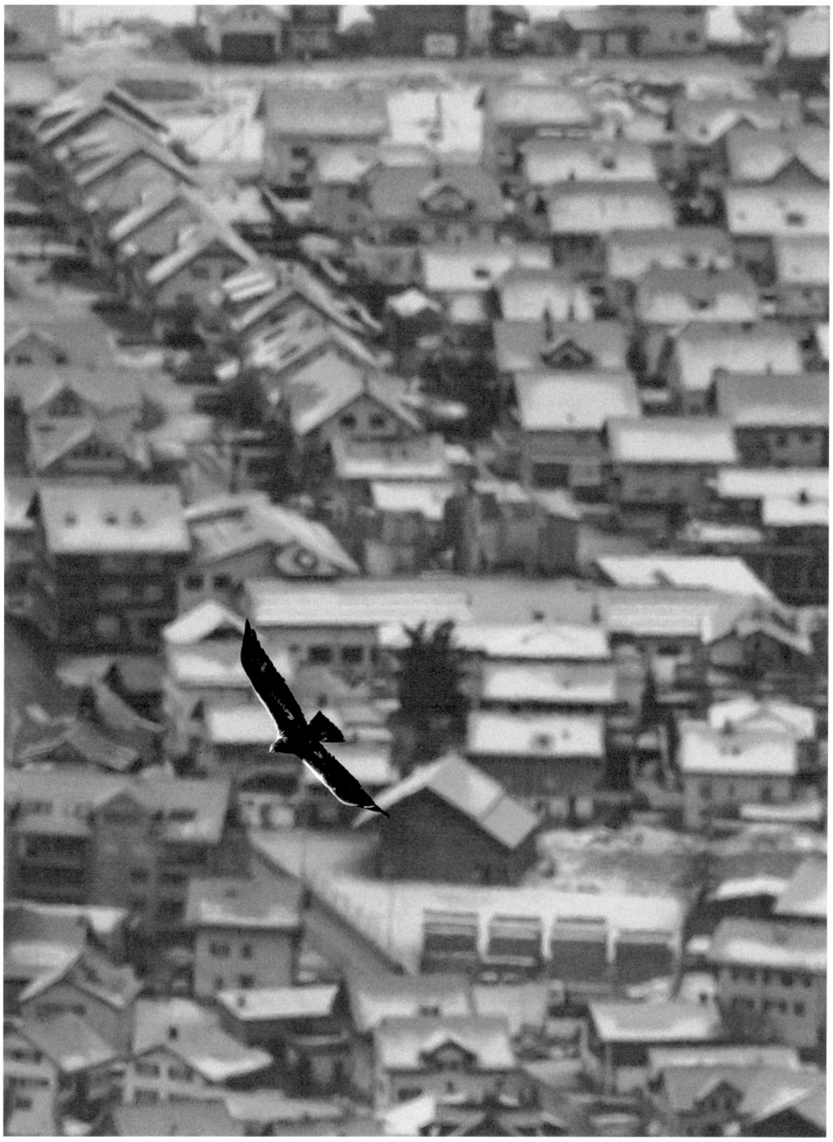

THE WARRIOR AND THE EAGLE

LEGEND FROM THE DELAWARE INDIANS \ 107 \

Long ago the Delawares believed that if a brave could pluck a feather from the tail of a live eagle and wear that feather, he would not only always be brave and of great courage, but good fortune would always follow him. Therefore young hunters used to try to catch eagles by putting pieces of wolf meat on high cliffs, eagles being very fond of wolf meat.

At one time there was a young brave who was very reckless, ambitious and daring. He wanted to get eagle feathers for a head dress and desired to pluck the feathers himself from live eagles, so he found a high place where eagles often came and baited the place for a few days with wolf meat. Then he killed a large wolf, took it to this place and hung a large piece of the flesh near the edge of the cliff. He then hid behind a big tree, with a forked stick, ready to capture an eagle.

Presently an eagle came to get the tempting morsel, but the young brave considered this eagle too small and drove it away. Soon another came, but this one also did not seem to suit the brave. He drove away several others, not being satisfied with the plumage of any of them. All at once he heard the flapping of heavy wings and there alighted before him an eagle much larger than himself. This eagle instead of looking like the others had red feathers, as if dyed in blood. This eagle did not take the wolf meat, but came straight to the brave, seized him in his talons and carried him away to a high cliff, from which it was impossible to escape, except by jumping down, which would have been certain destruction.

On this cliff was a large nest containing four young eagles. The large eagle left the brave in the nest and said to him: "You shall stay here and care for my young until they are large enough to carry you back to where I got you. I am the head chief of the eagles. Your greed and ambition have brought you to this. You were not satisfied with the plumage of the birds I sent you. Now you shall stay here and suffer for your greed, and perhaps when you return you will be glad to take such feathers as we give you." There was nothing else for the young brave to do but stay and guard the young eagles, and this he did so well as to win the friendship and love of the young eagles as well as the old eagle, who occasionally came to the nest, bringing in his talons a deer, rabbit or other game.

Finally, after the brave had been there many days and the young eagles had learned to fly, they would sometimes be away nearly all day and leave him alone. He would get very lonely and wonder if they were going to leave him to die of starvation or eat him up, or whether they really meant to take him back where the old eagle found him. He was not kept long in suspense, however, for one day the large eagle came again and said: "Now, my young friend, my grandchildren here shall carry you back to where I found you. I will go along to see that they do not drop you until you reach the place in safety." Accordingly two of the young eagles seized the brave in their talons and flew toward the cliff where he had been tempting the eagles with wolf meat. It was not far from the nest and they soon reached the place in safety. There the brave found some eagle feathers which he was glad enough to take without plucking them from a live eagle, and he returned with them to his people.

The lesson he learned from his adventure is that opportunities will finally cease to come if you continue to brush them aside, hoping for a better one.

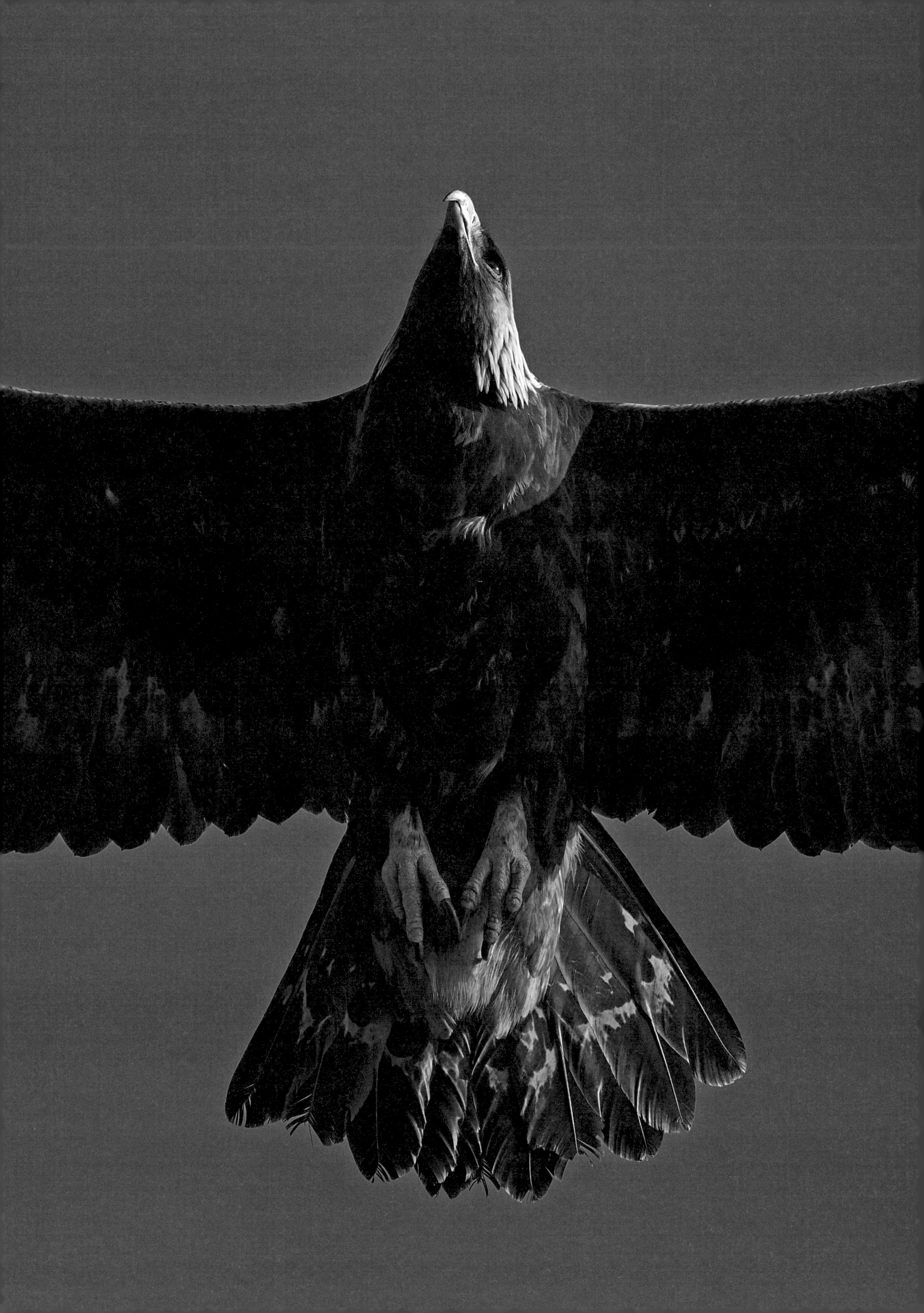

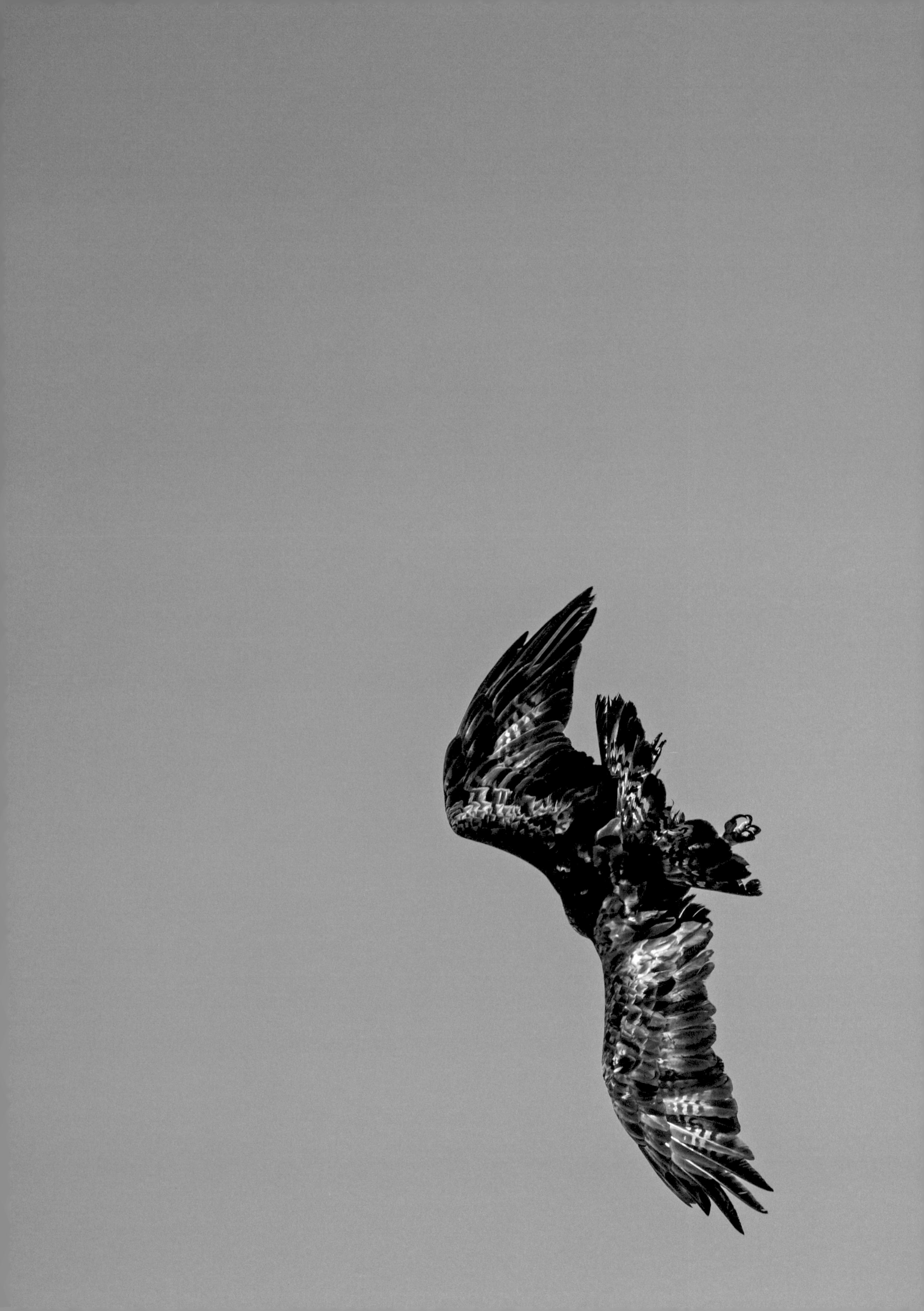

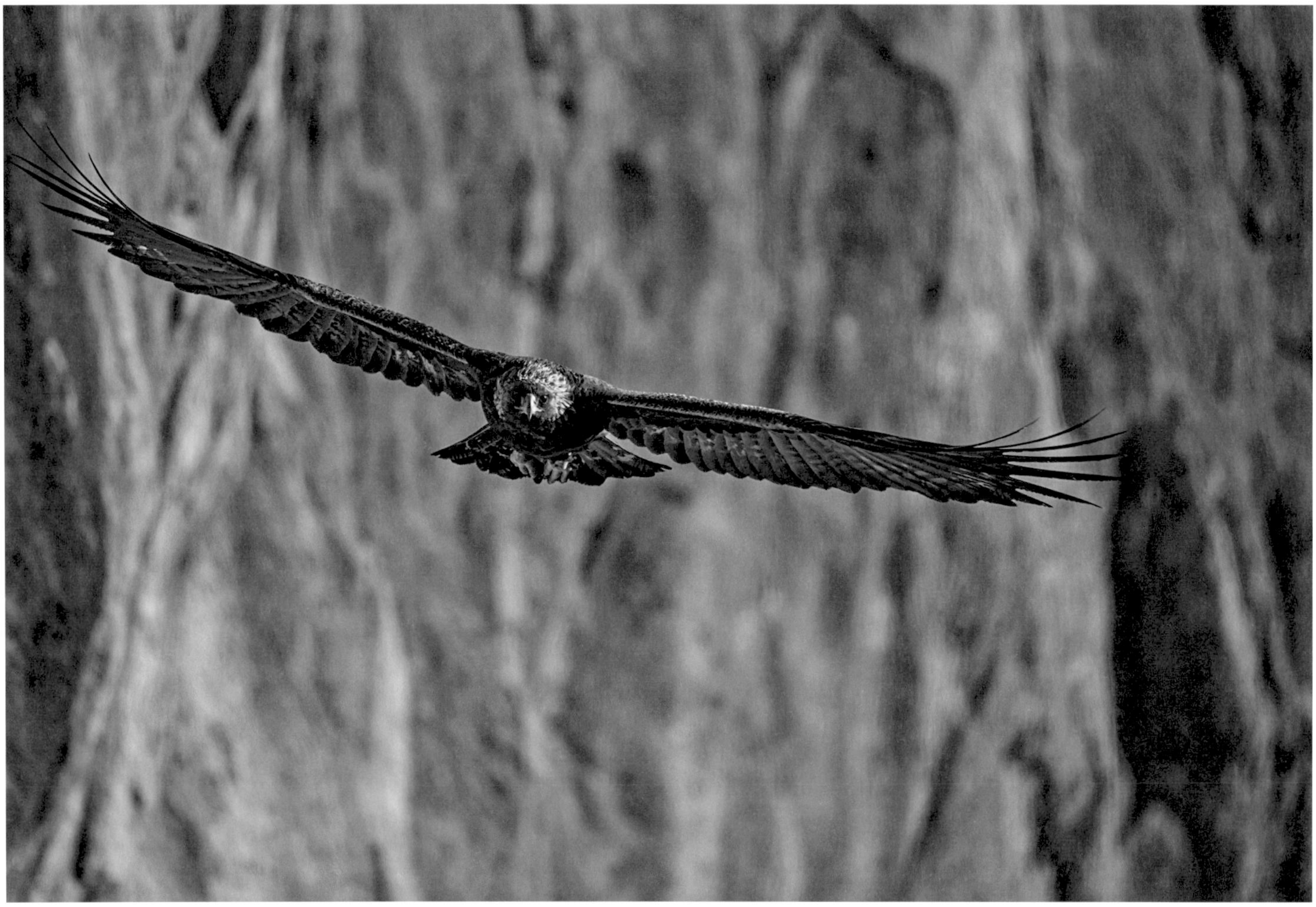

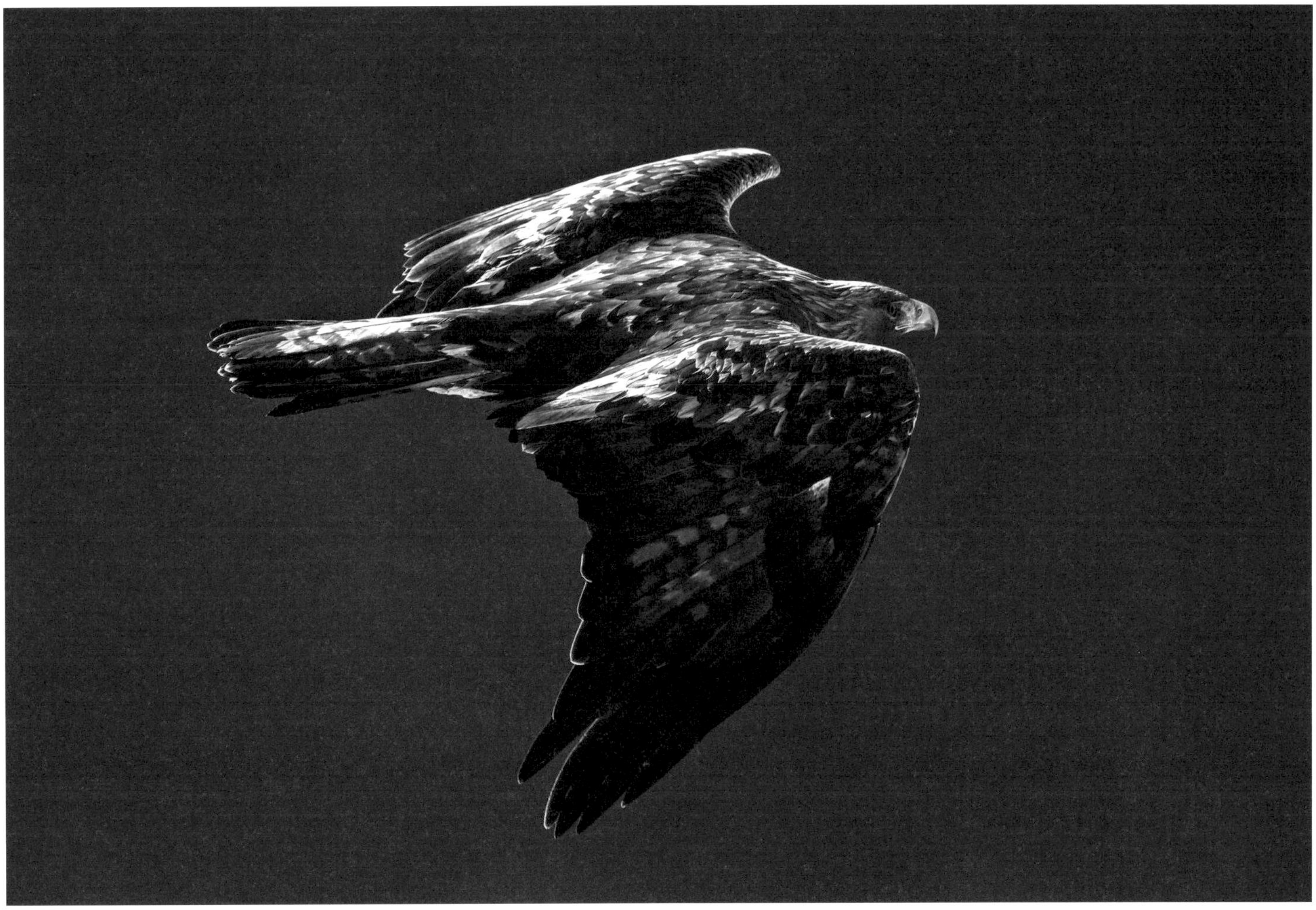

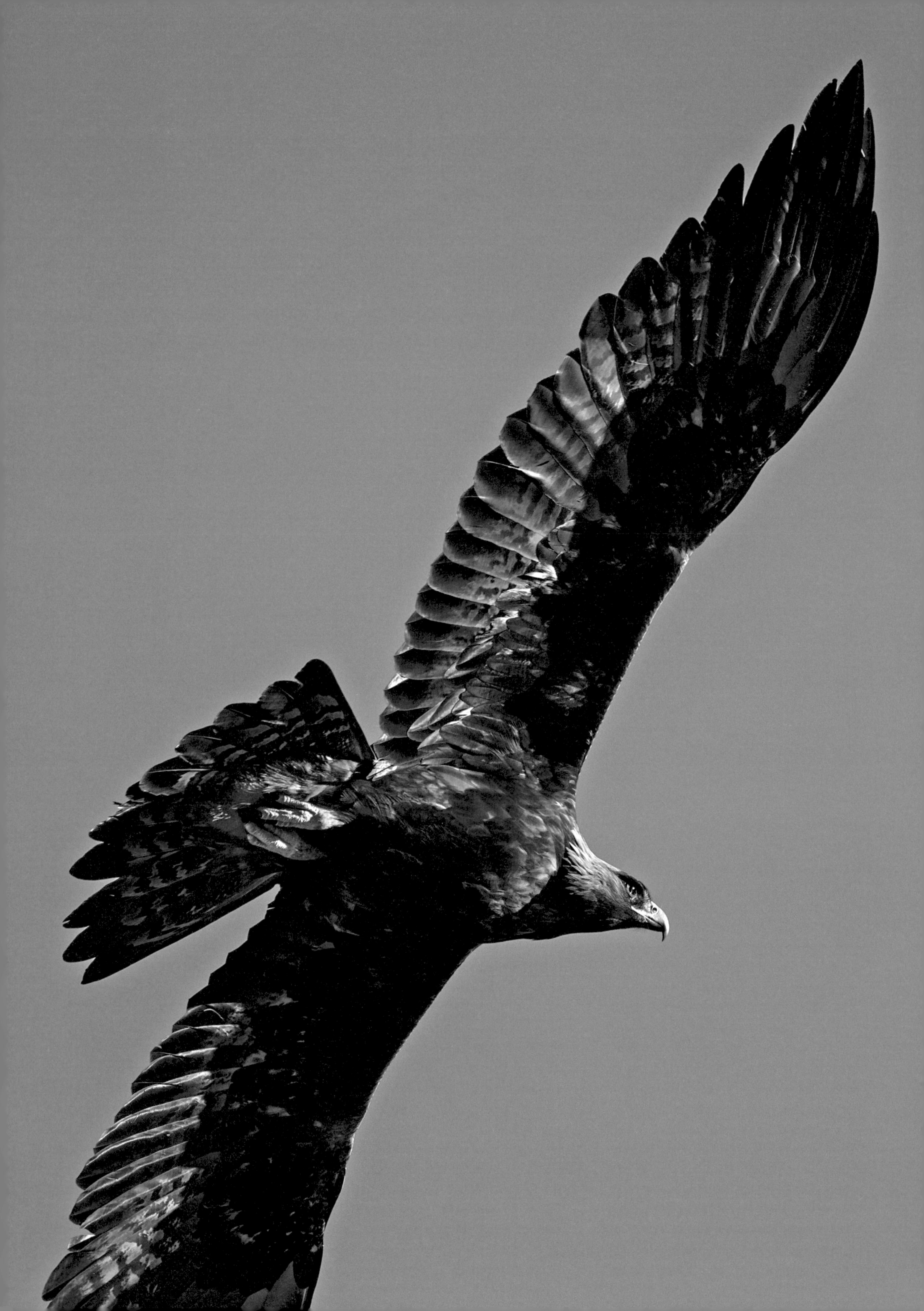

THE GIFTED OF HEAVEN

FRANCES FULLER VICTOR \
1826 — 1902 \ 108 \

Song of the Eagle

I'm the child of light, yet the darkest night
No terrors hath for me,
For the storm I ride, in a monarch's pride,
Or skim o'er the heaving sea.
When lowering clouds, like sable shrouds,
Wrap the earth in deepest gloom,
I join the surge in the funeral dirge,
O'er the sailor's watery tomb.

And I love to rest on the summit crest
Of the proudest mountain's height,
While the clouds below lie like wreaths of snow,
Yielding homage to my might.
In my pride I go where eternal snow
Has crested the mountain's brow,
And laugh at the storm, and the blackened form
Of the threatening clouds below.

Mid the lightning's flash, and the thunder's crash,
I scream for my own delight,
For I love to hear, so loud and clear,
My voice ringing out in the night.
Not so proud a one ever gazed on the sun
As the eagle bird, I trow,
Stooping to rest on the towering crest
Of the highest mountain's brow.

In the pride of a king, with folded wing,
I gaze on ruined Tyre;
By Heaven's decree it was given to me,
And no power to give is higher.
From land or sea God hath chosen me,
And a favored bird am I –
The gifted of Heaven, to whom power is given
Over earth, and sea, and sky.

I care not for earth, though I had my birth
On the proudest height she owns;
And I'd rather ride o'er old ocean's tide
Than sit on her rocky thrones.
But I love the sun, and could I have won
A home in its realms of light,
With a laugh of scorn from this earth I'd turn,
And soar to my home in delight.

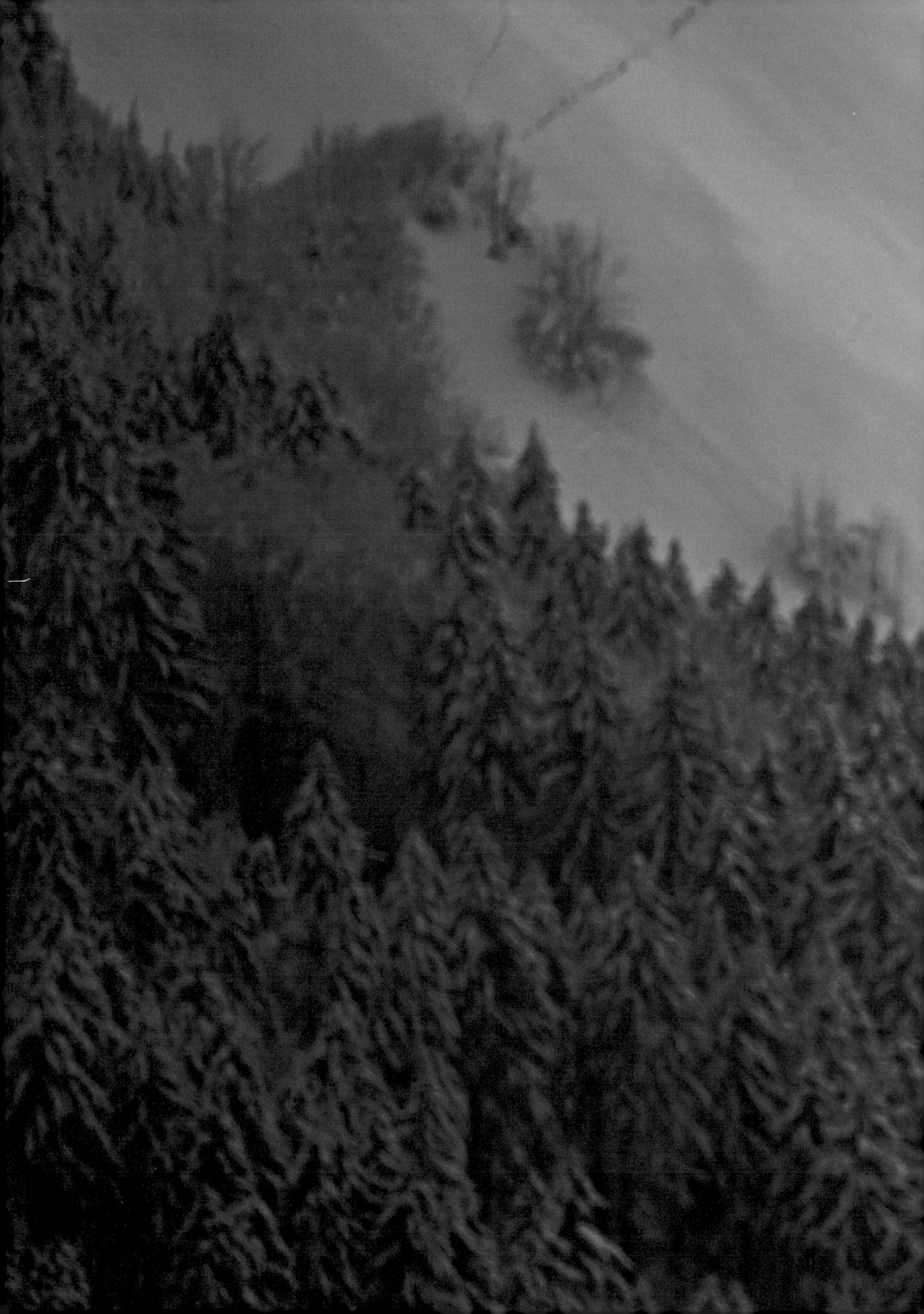

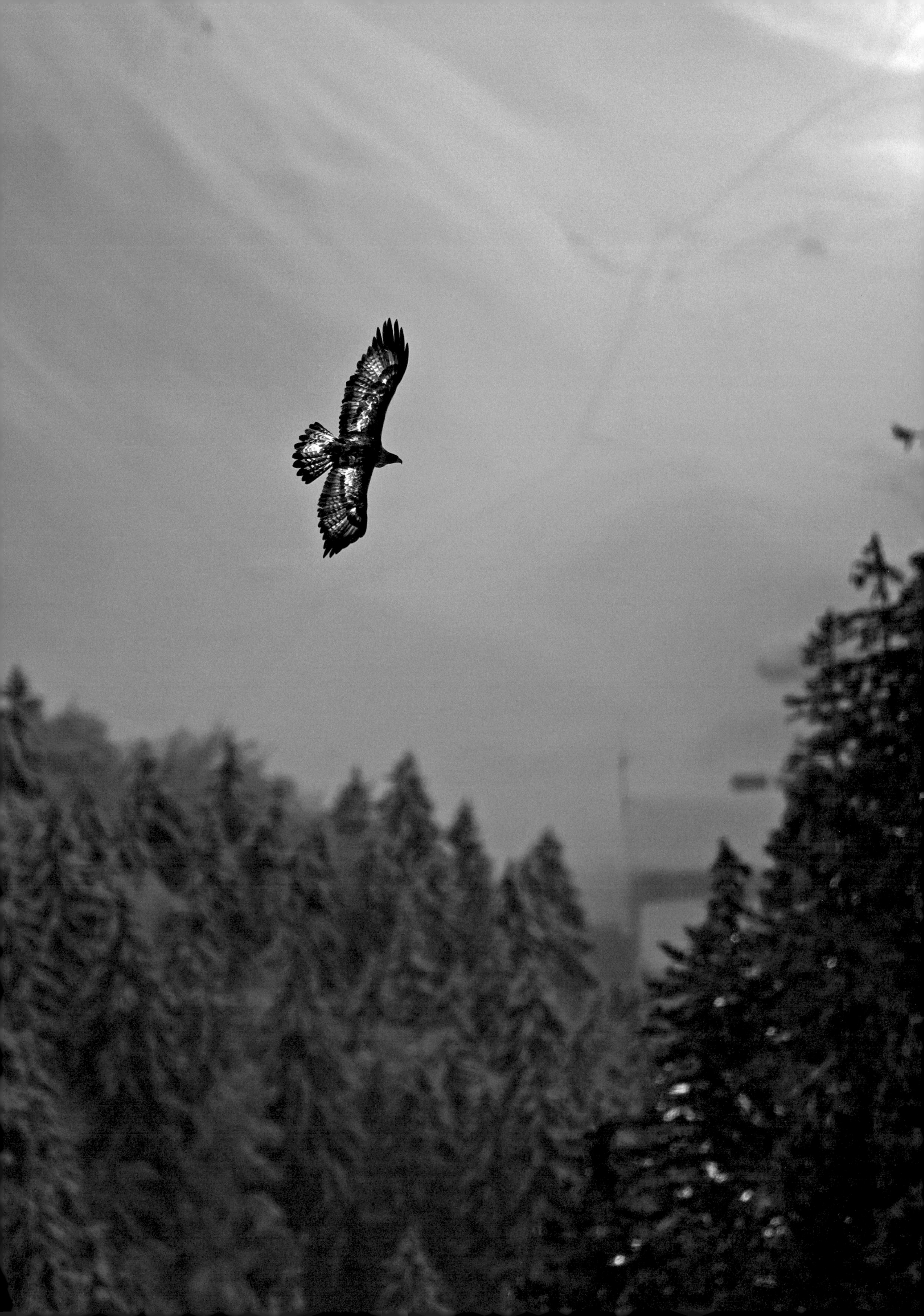

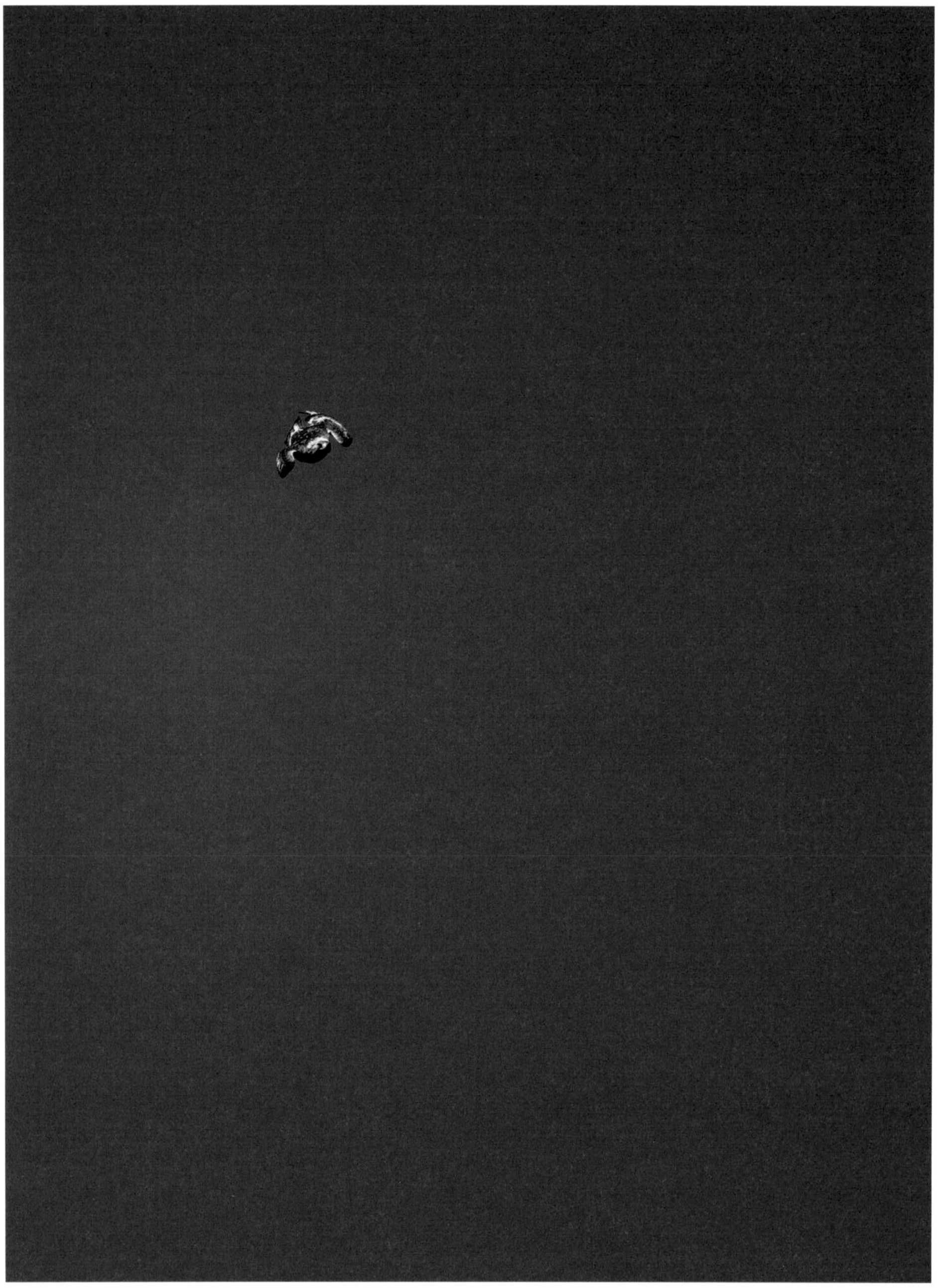

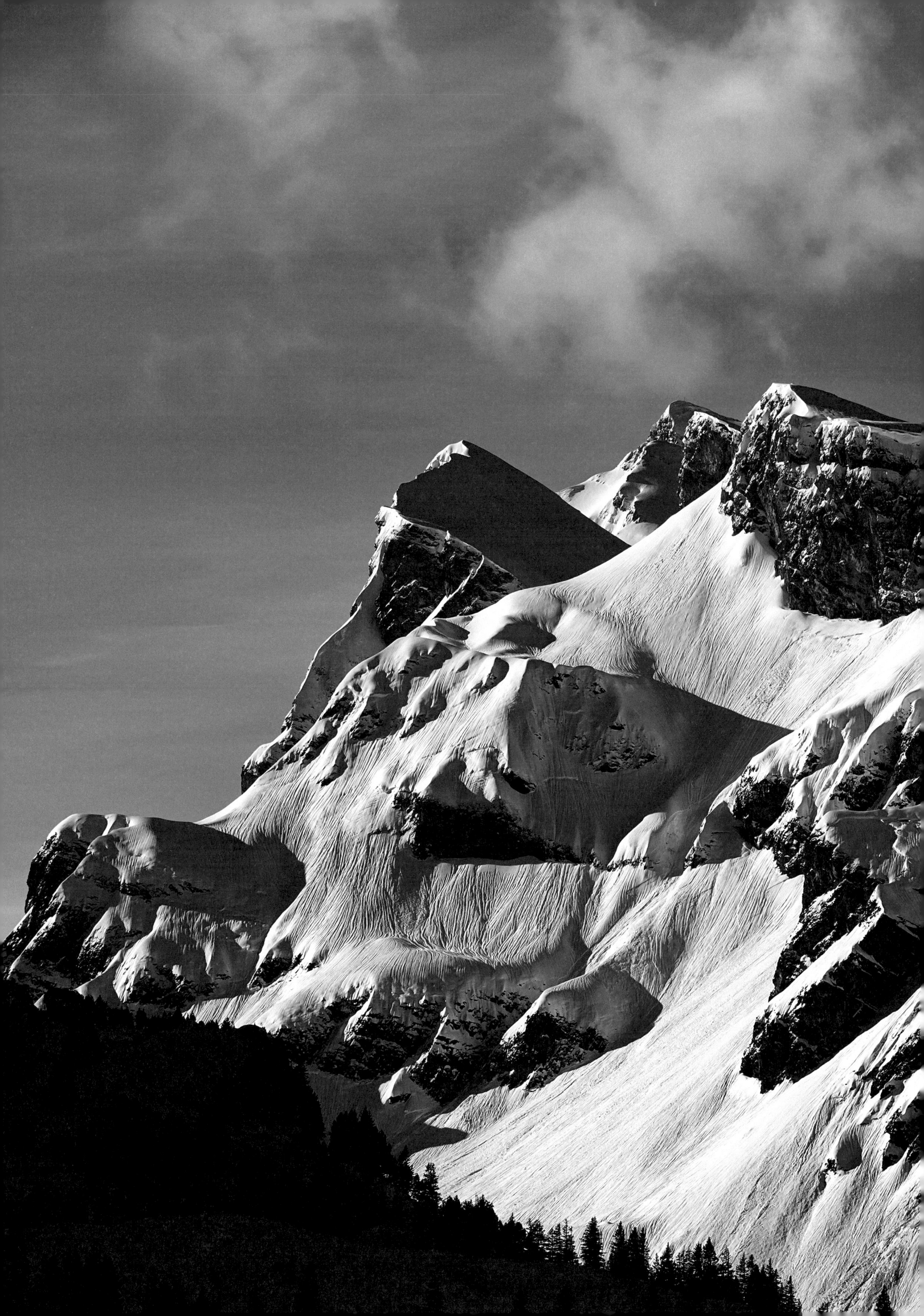

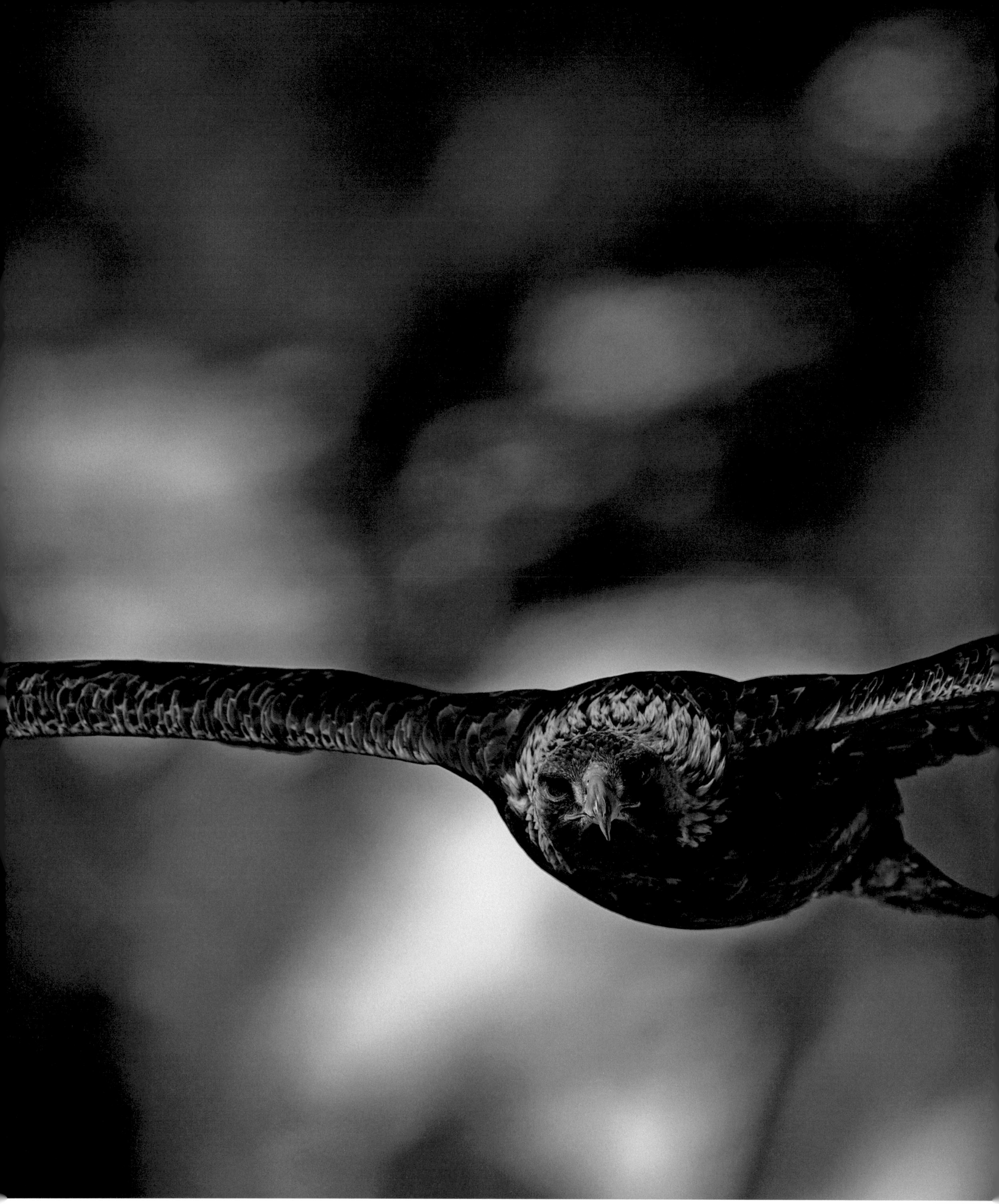

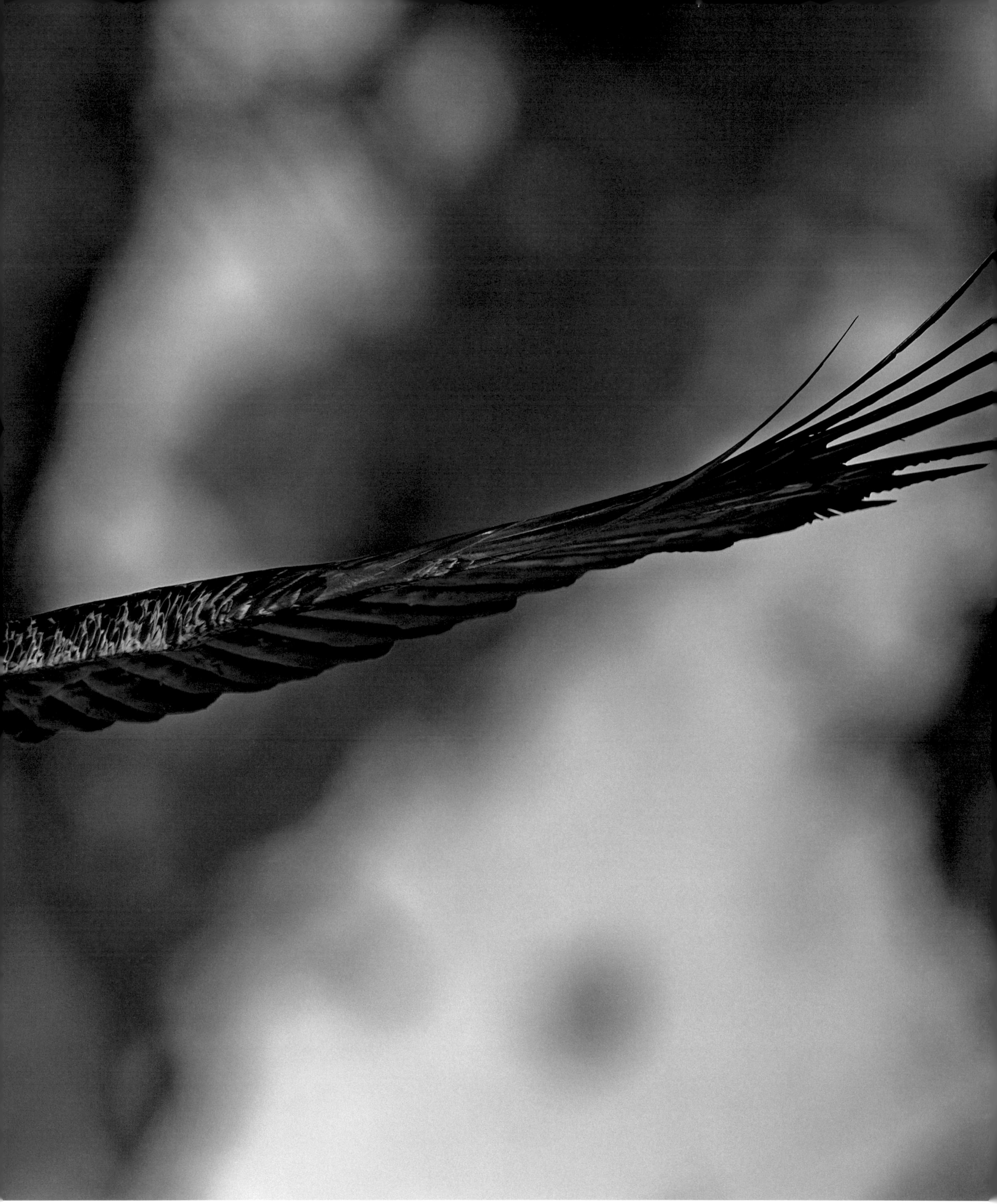

THE WISDOM OF AN EAGLE

JOHN FIRE LAME DEER
(TȞÁȞČA HUŠTÉ IN LAKOTA) \
1903 — 1976 \ 109 \

In an eagle there is all the wisdom of the world.

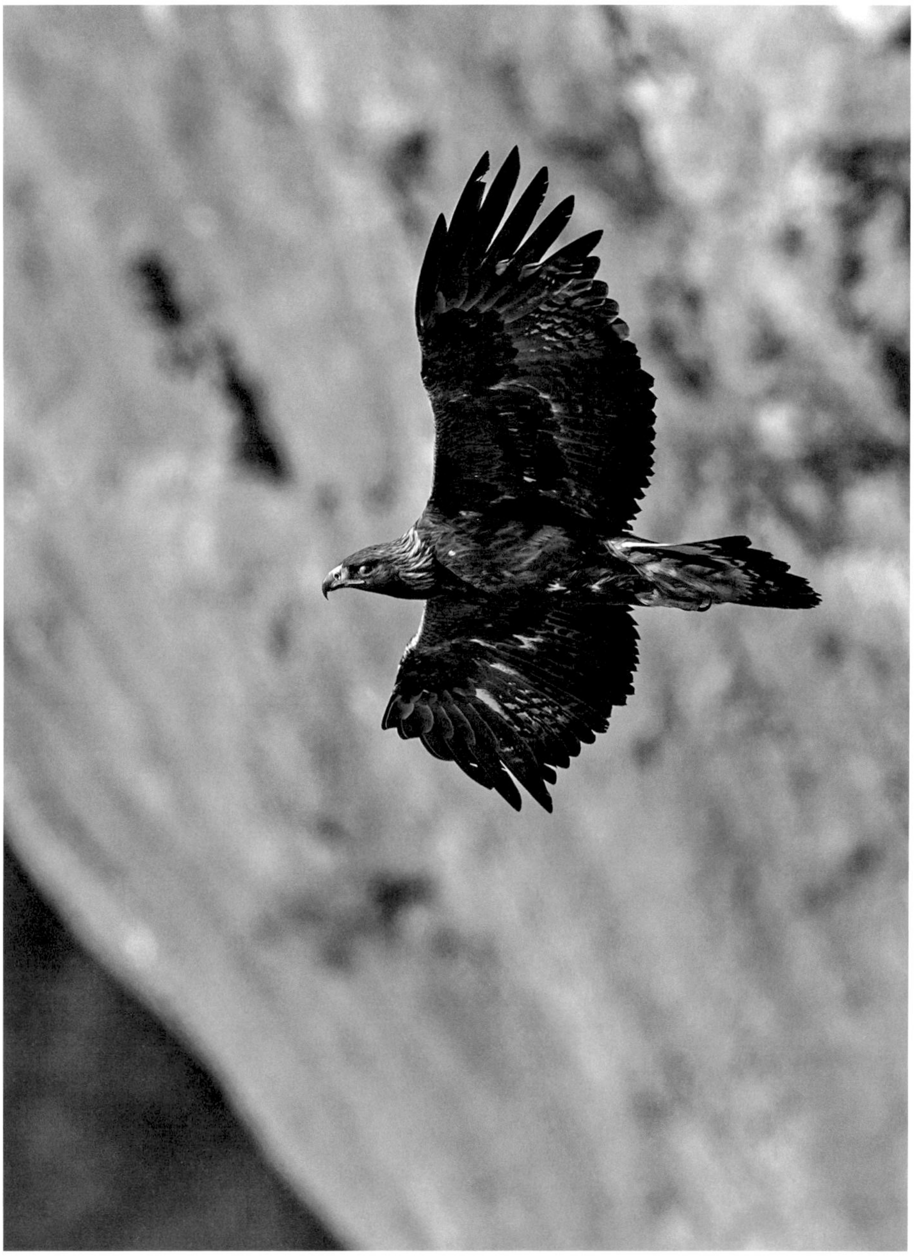

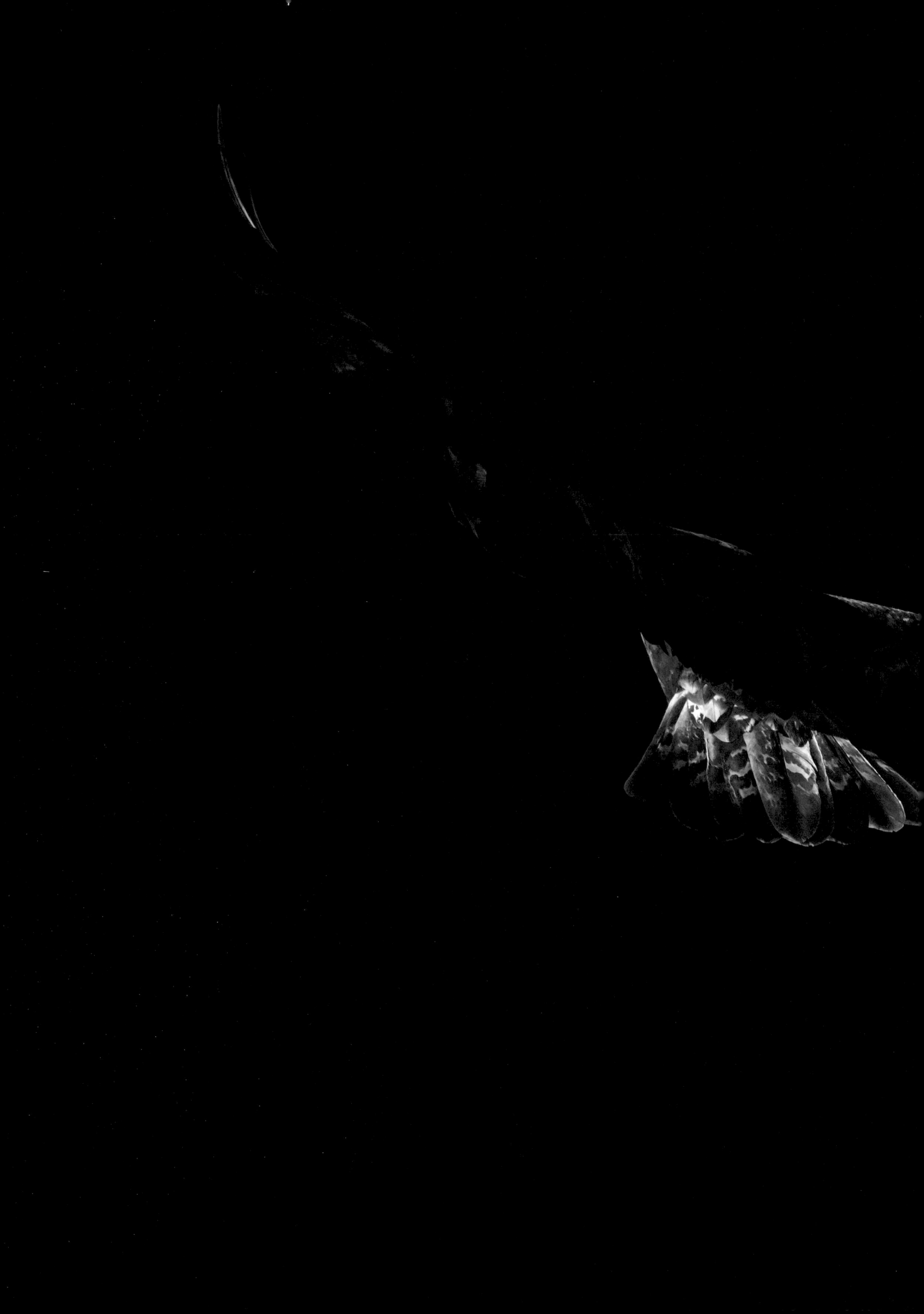

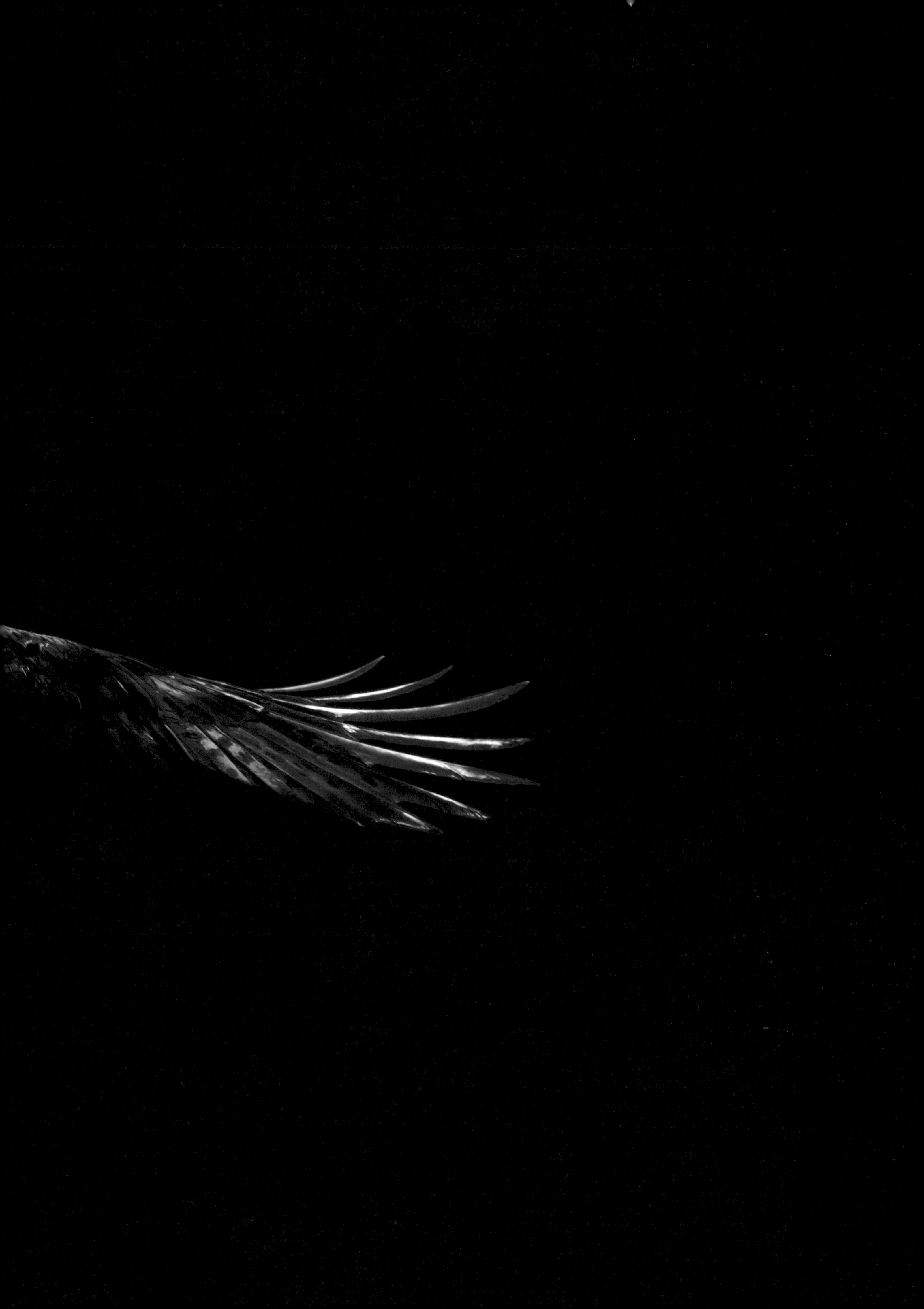

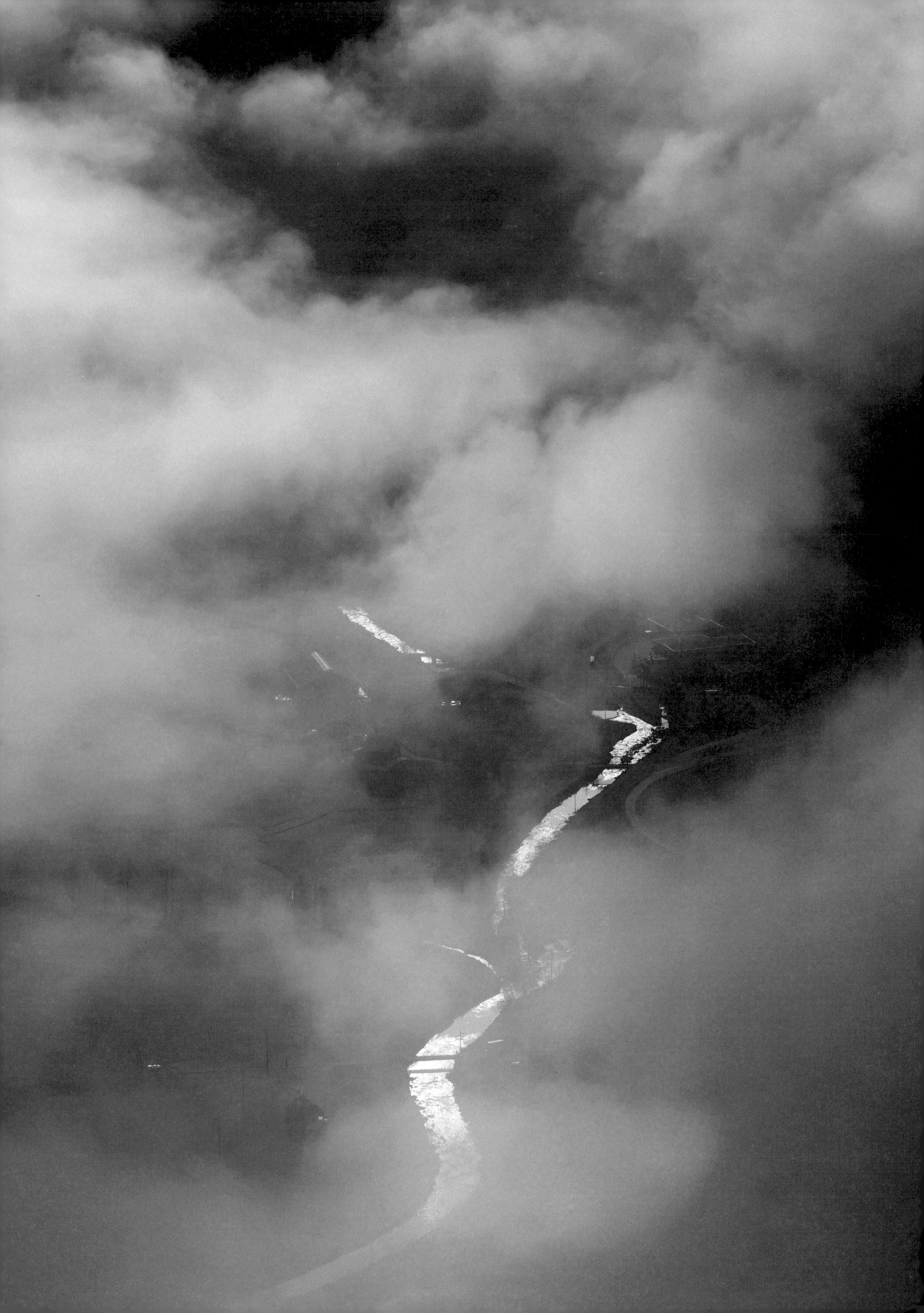

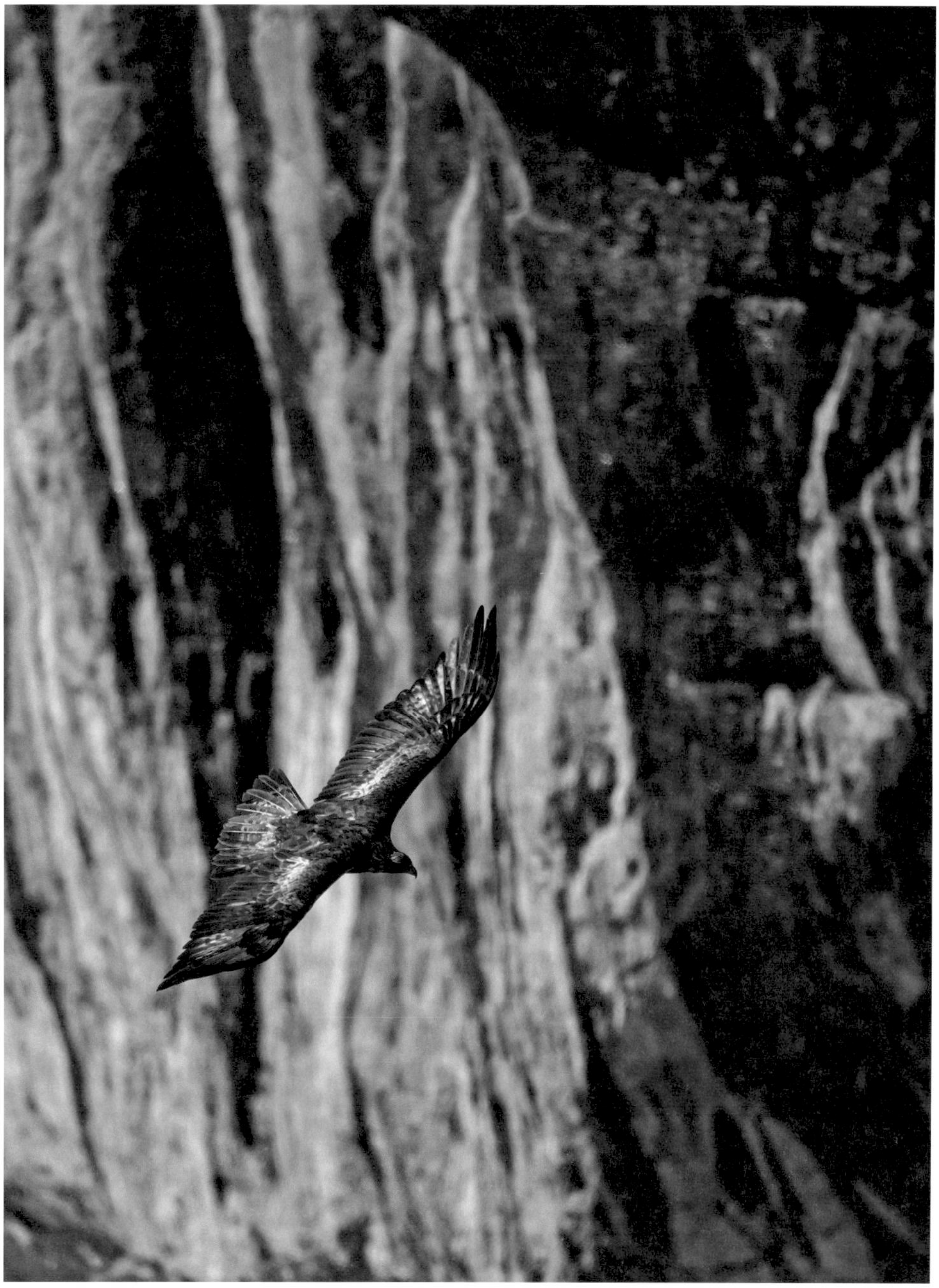

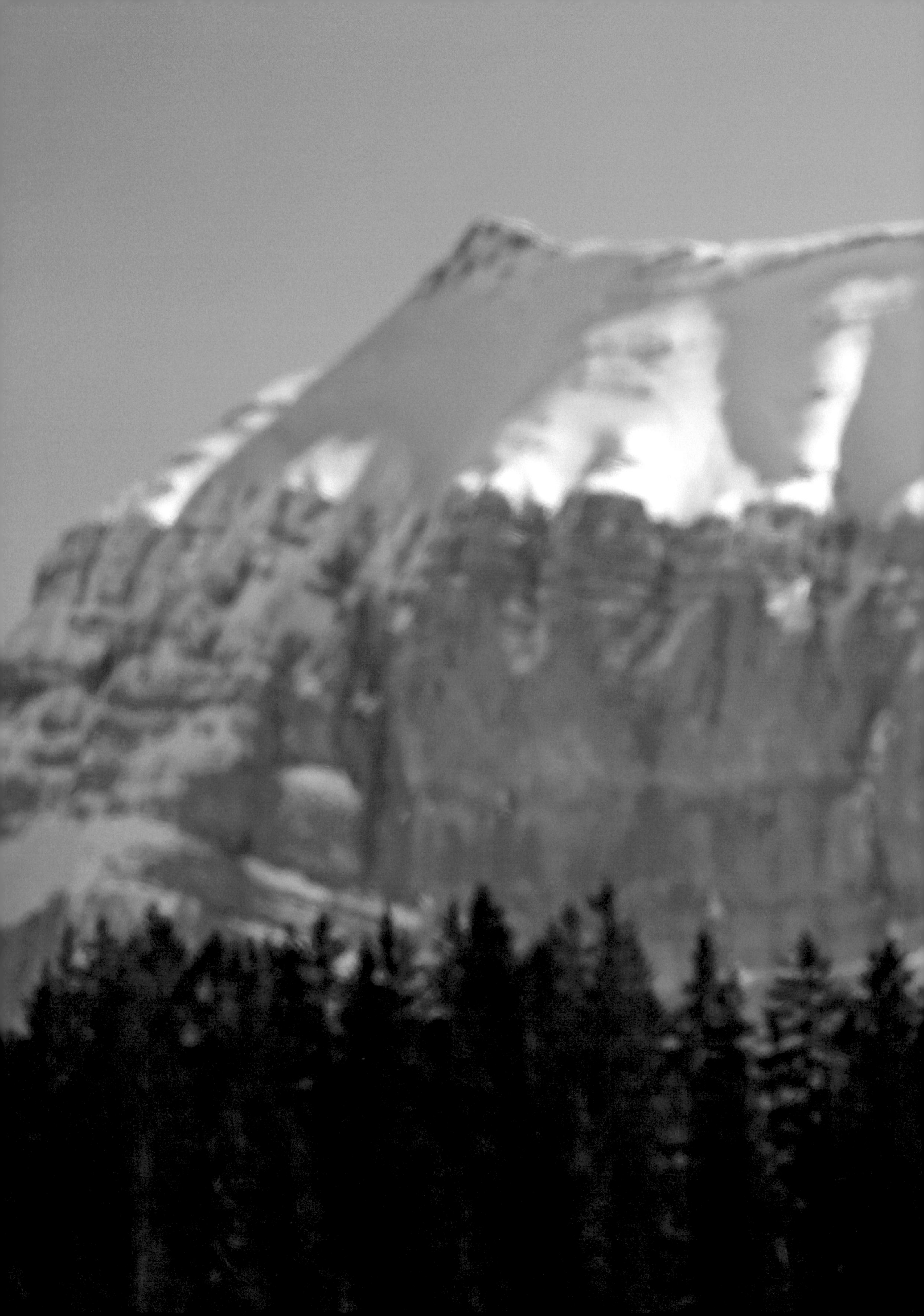

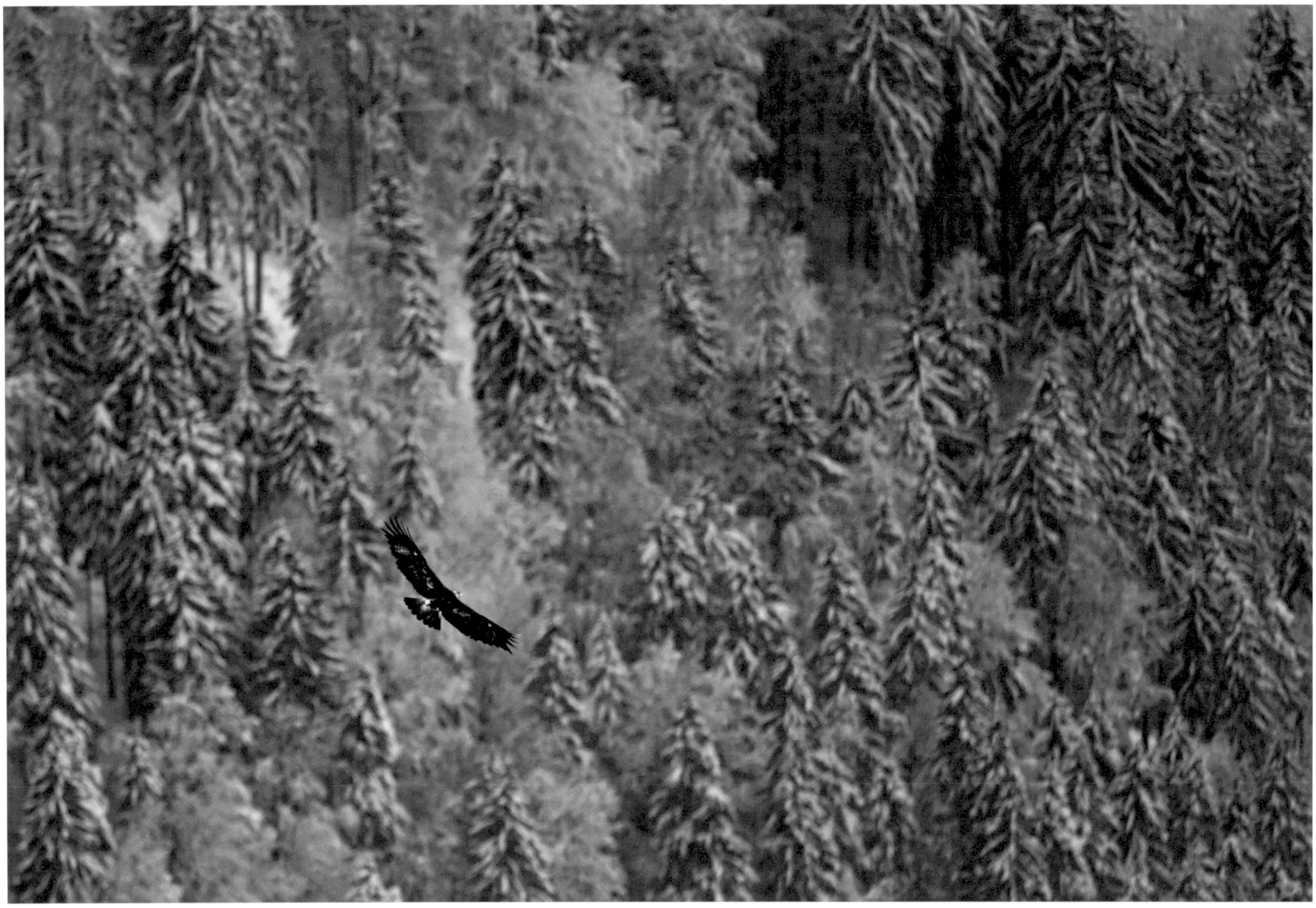

THE GIFT OF THE FAIRIES

GOTTHOLD EPHRAIM LESSING \
1729 — 1781
TRANSLATED AND SLIGHTLY MODIFIED \ 110 \

Two charitable fairies appeared at the cradle of a young prince, who later became one of the most powerful kings of the country.

"I herewith endow my beloved prince with the sharp-sighted vision of the eagle, who does not even miss the smallest fly in his large empire." The first announced.

"This is a really beautiful gift." The second fairy commented. "The prince will certainly become an understanding monarch who will be loved by his people. The eagle has not only the sharp-sighted vision to detect even the smallest fly, but also the noble contempt not to chase after them. And this will be my gift for the young prince!" She added.

"Thank you, sister, for this wise constraint." The first fairy replied. "It is true, many would have been much greater kings had they not used their piercing mind to try to understand even the smallest matter."

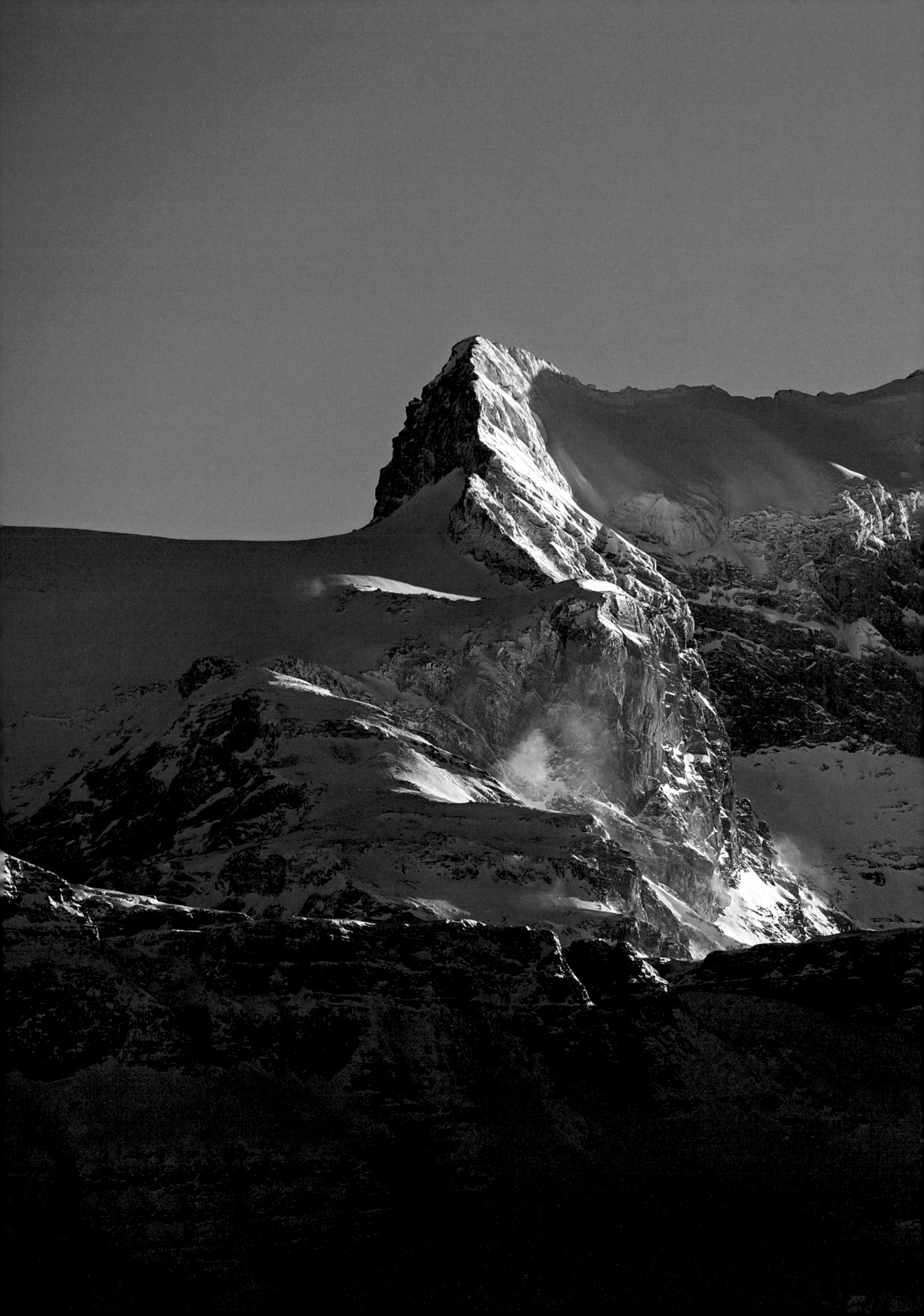

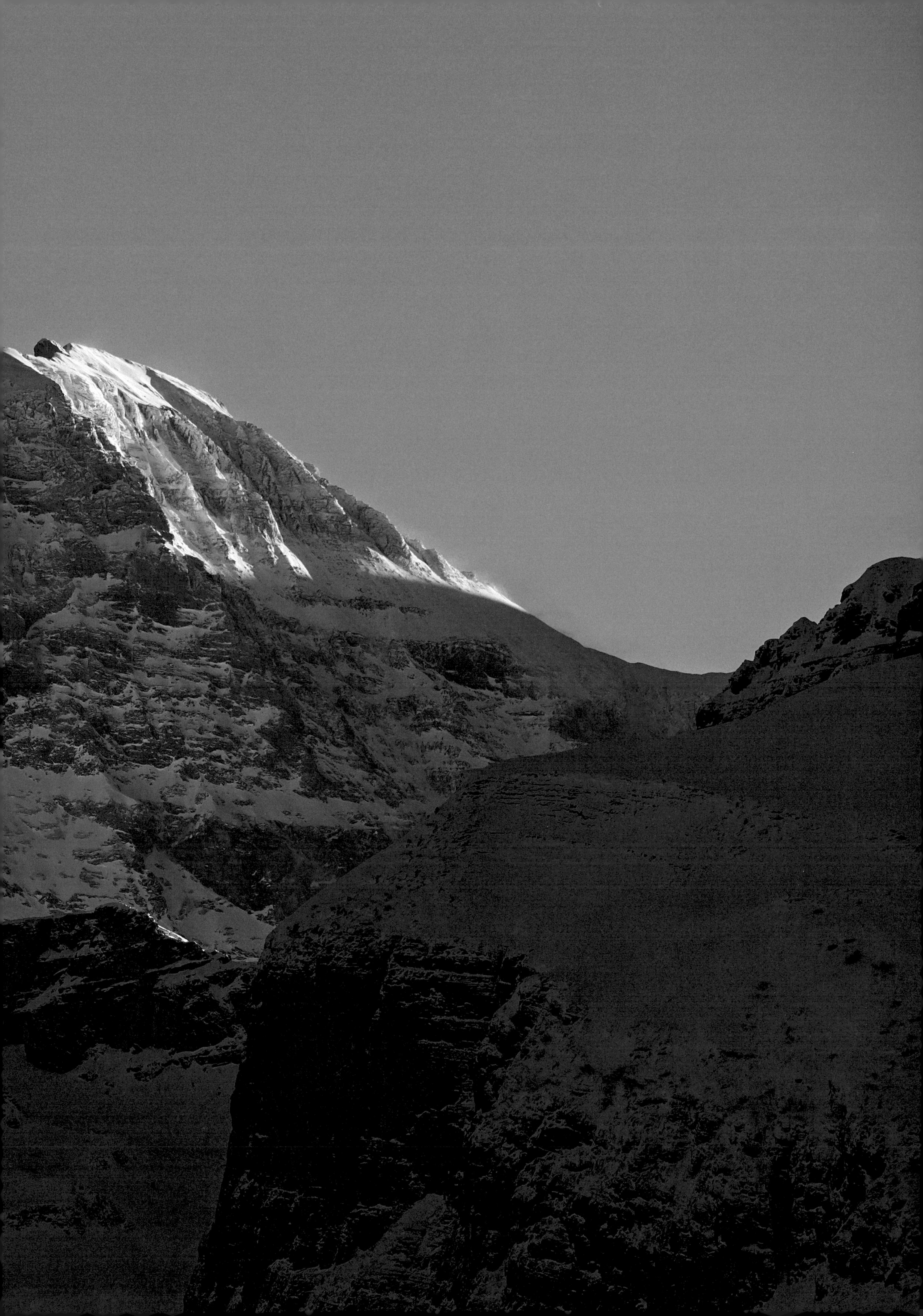

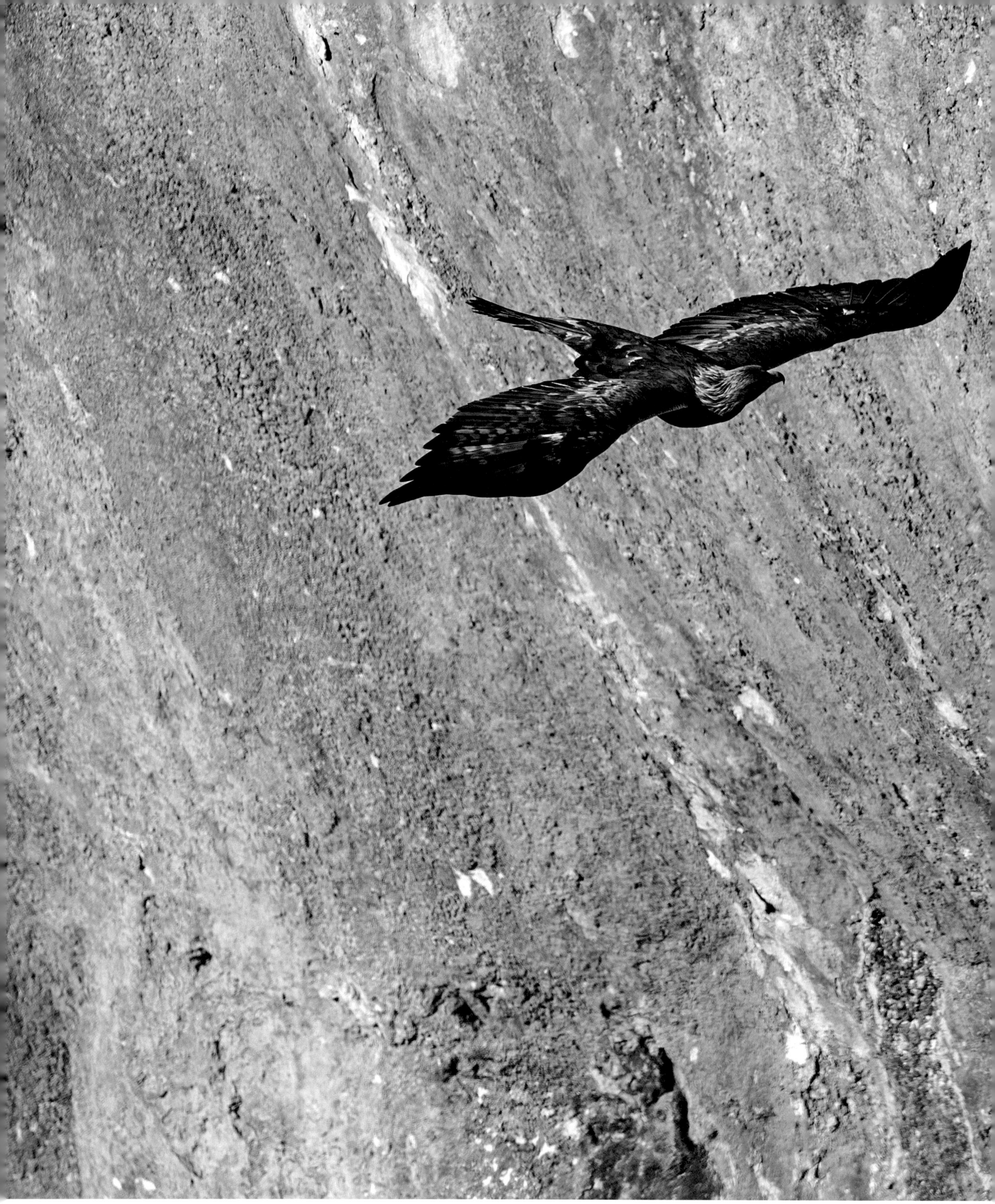

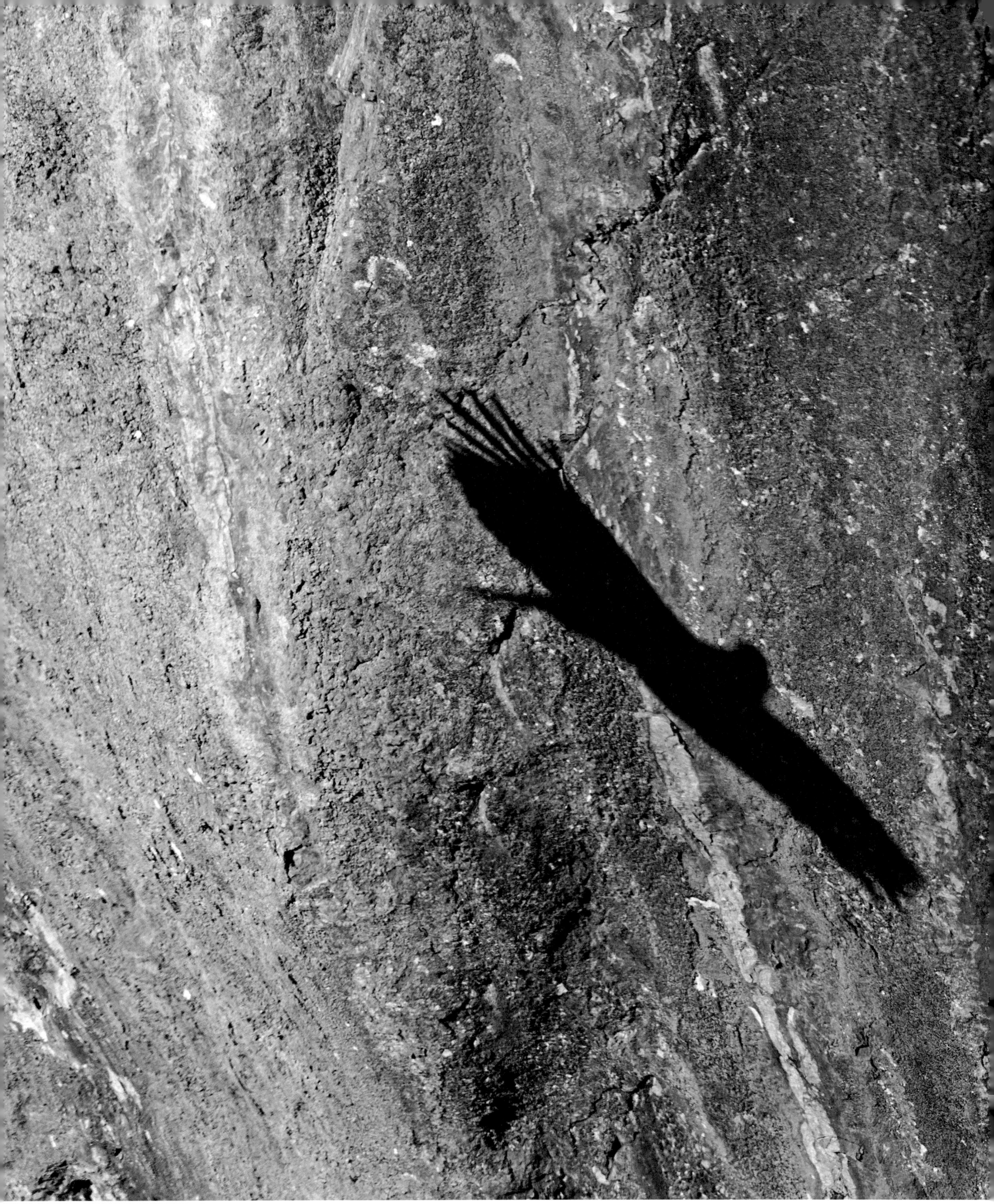

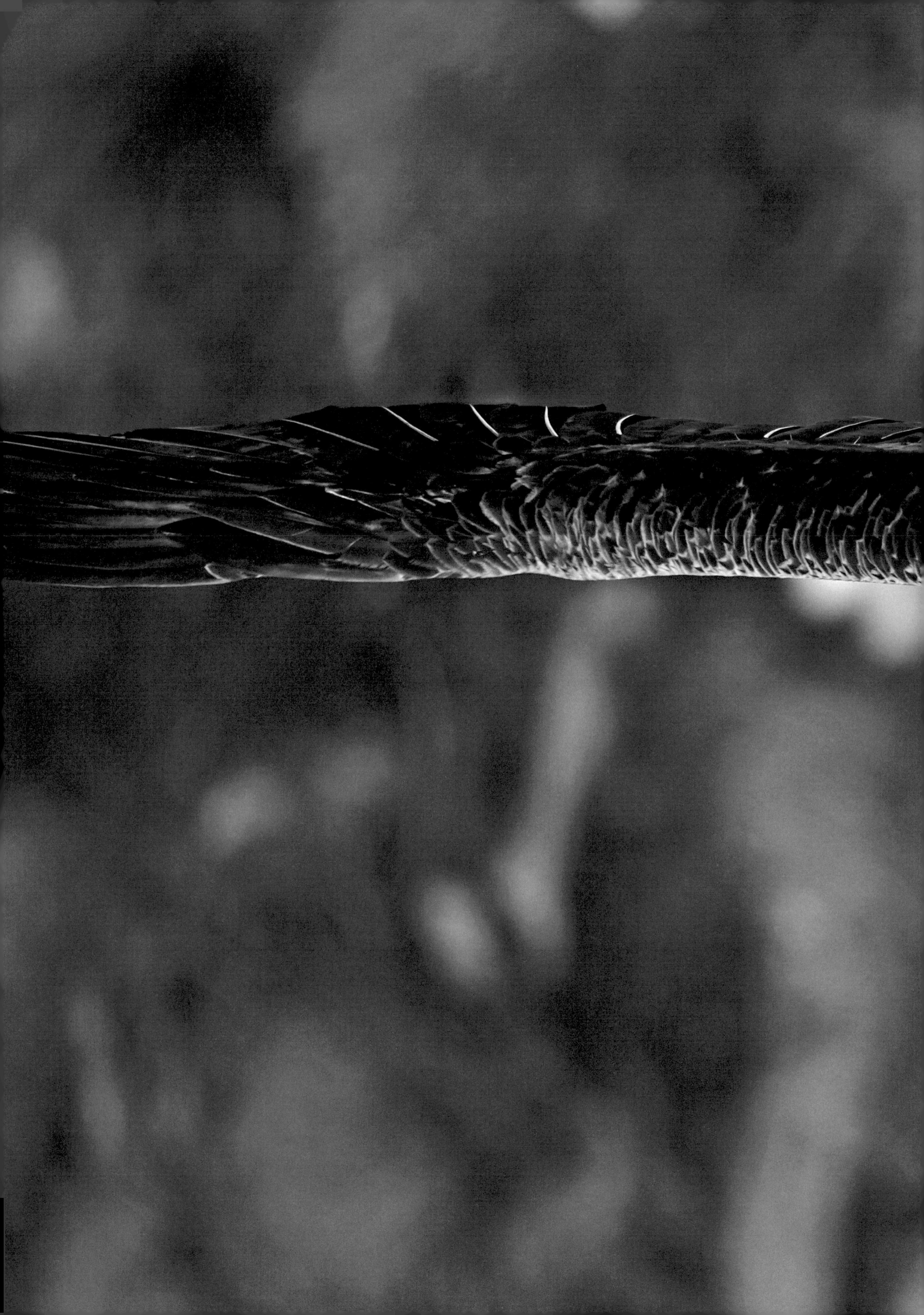

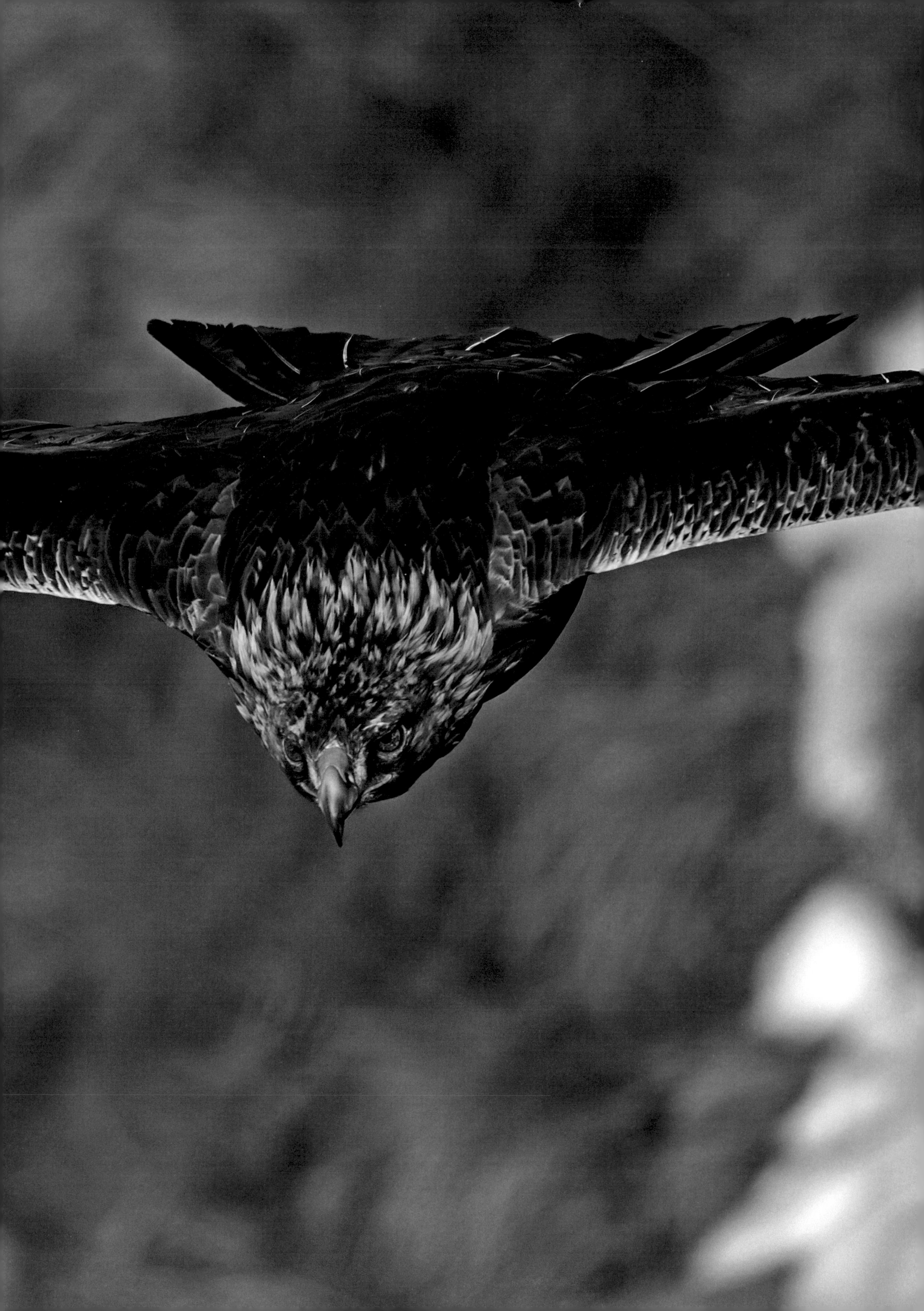

SOARING LIKE AN EAGLE

AZTEC POEM
DEDICATED TO WOMEN
THAT GAVE BIRTH \ III \

O my daughter, O valiant woman, you worked, you toiled.
You soared like an eagle, you sprang like a jaguar,
you put all your strength behind the shield, behind the buckler; you endured.
You went forth into battle, you emulated our mother, Cihuacoatl Quilaztli,
and now our lord has seated you on the Eagle Mat, the Jaguar Mat.
You have spent yourself, O my daughter, now be tranquil.
What does our lord, Tloque Nahuaque will?
Shall he bestow his favors upon each of you separately, in separate places?
Perhaps you shall go off, and leave the child that has arrived.
Perhaps, as small as he is, the creator will summon him, will call out to him,
or perhaps, he shall come to take you.
Do not be boastful of the child.
Do not consider yourself worthy of it.
Call out humbly to our lord, Tloque Nahuaque.

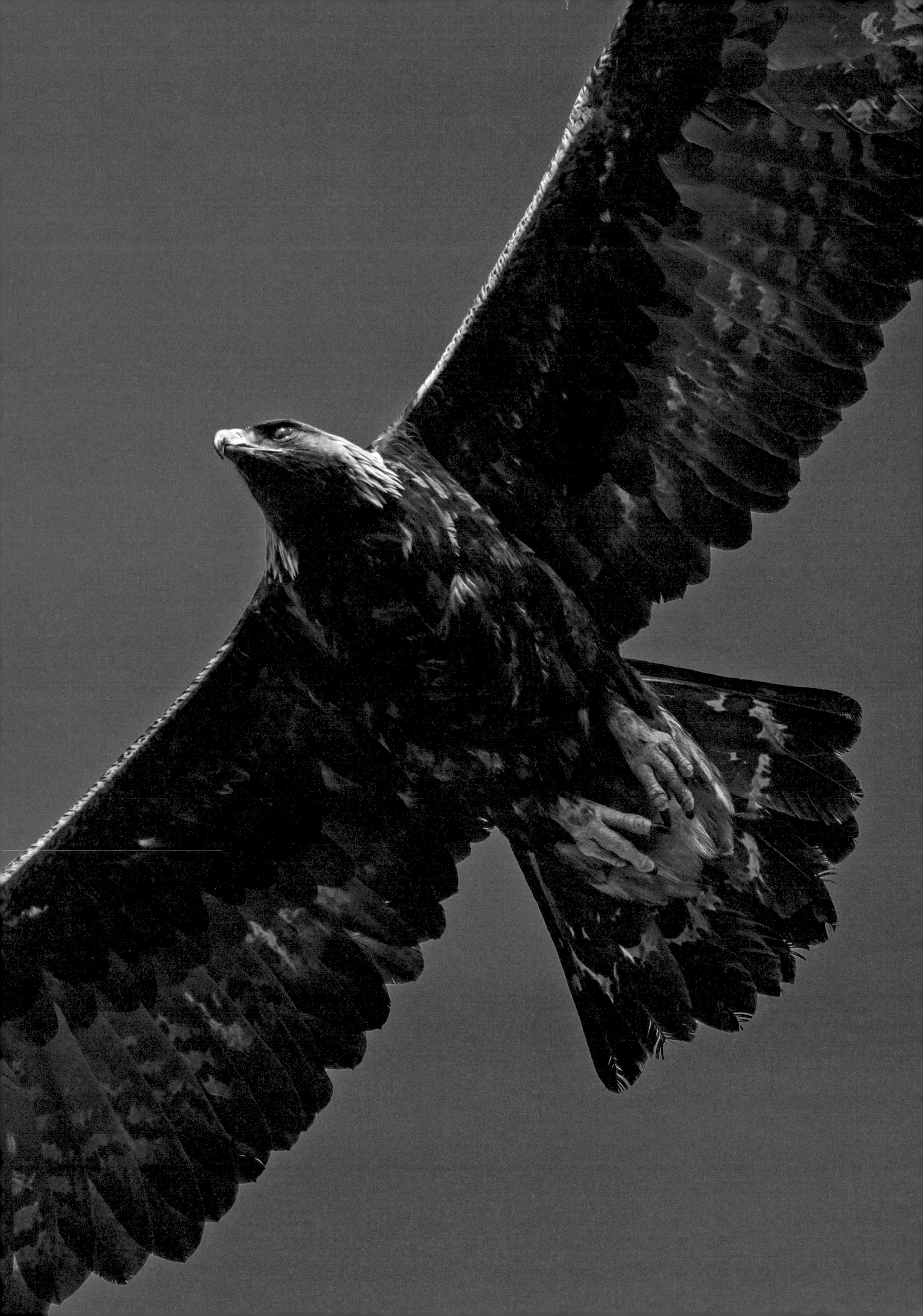

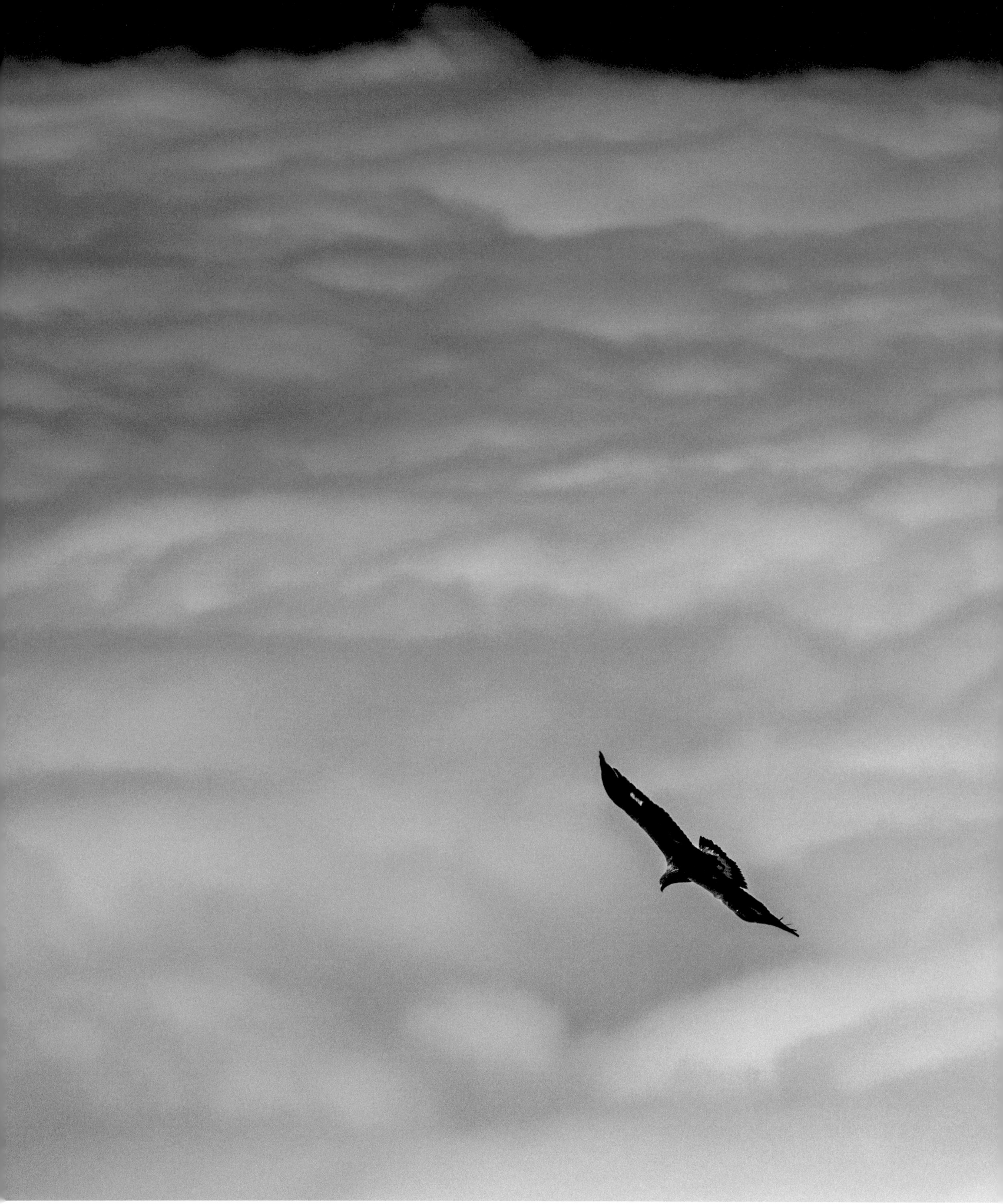

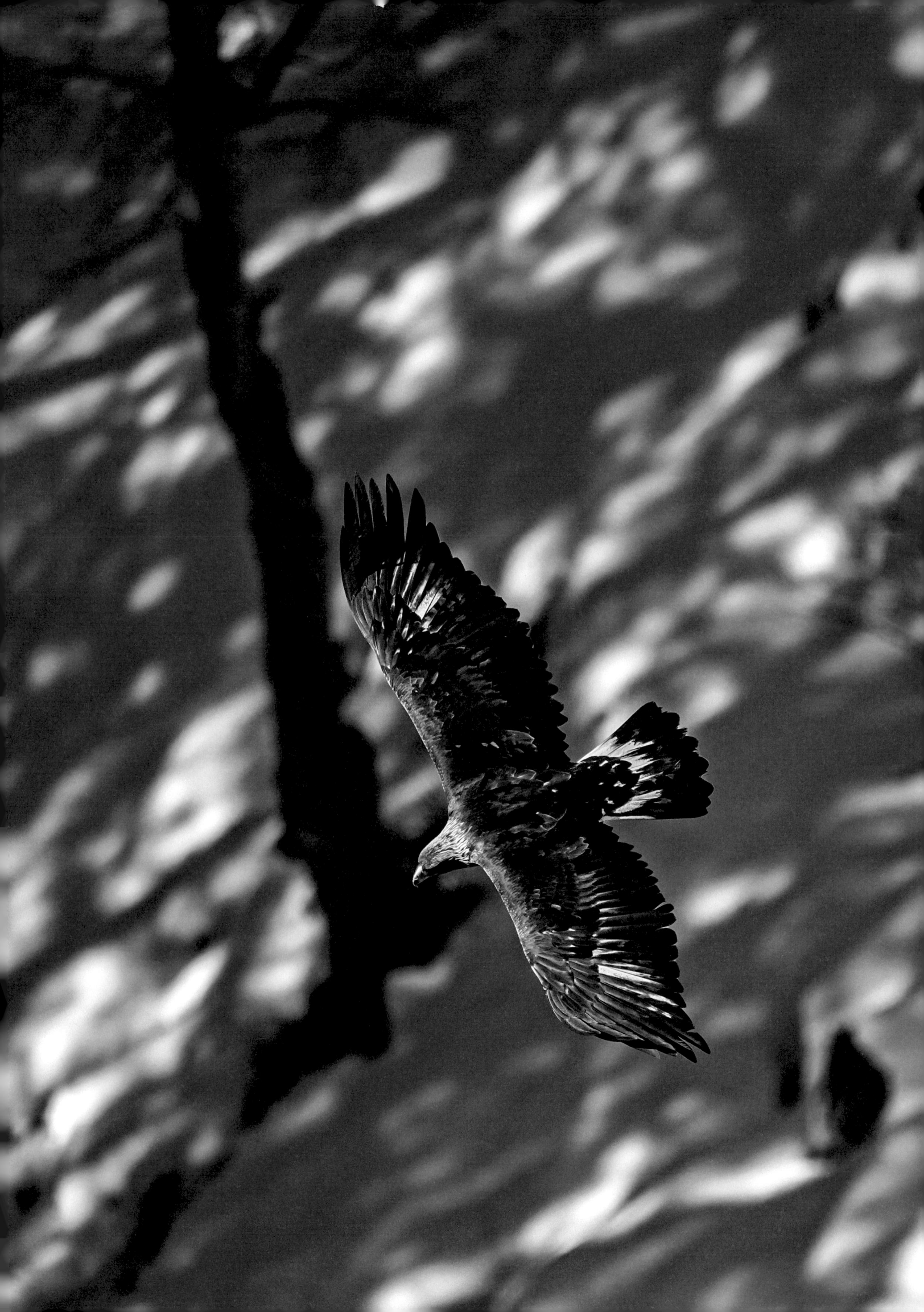

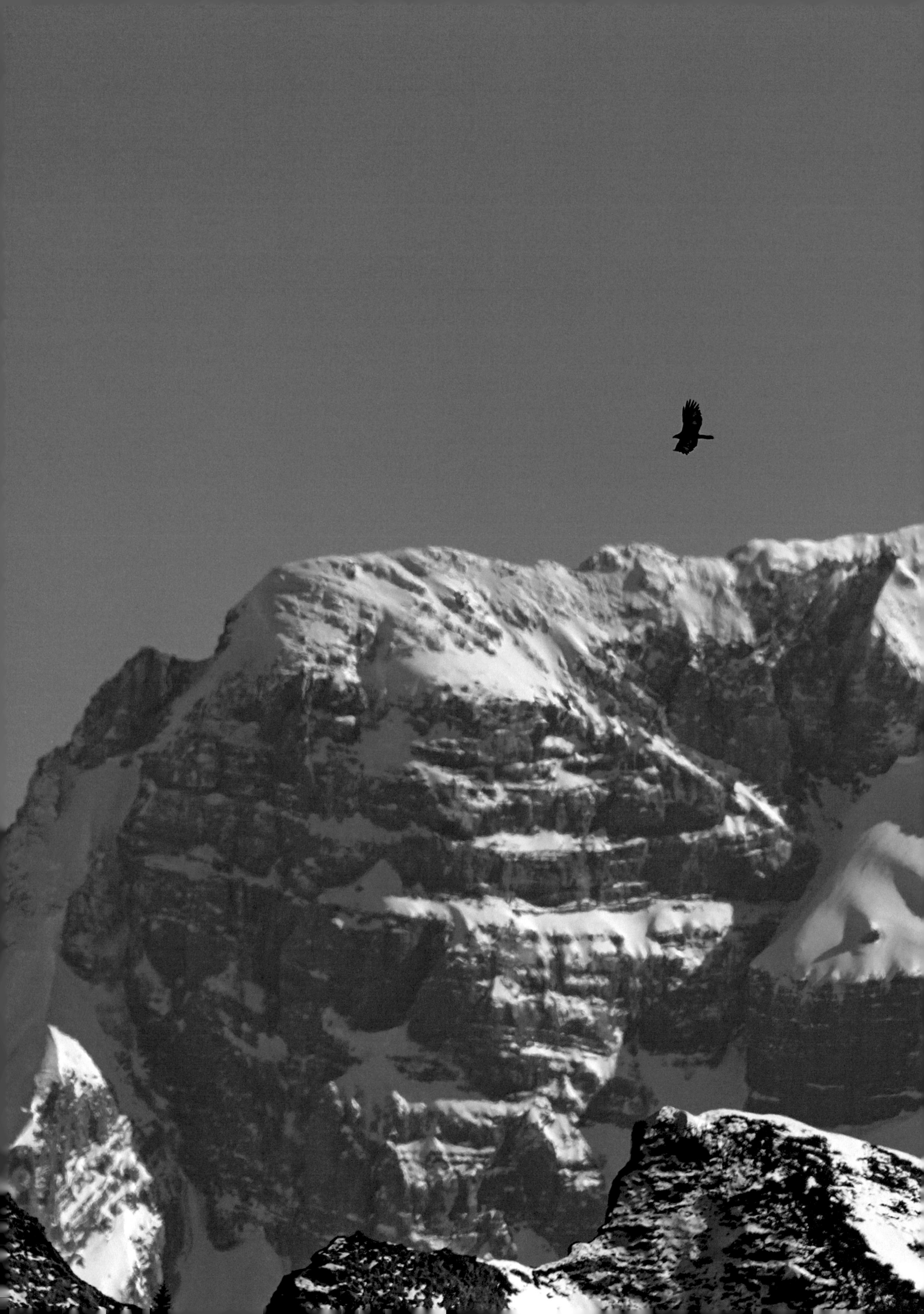

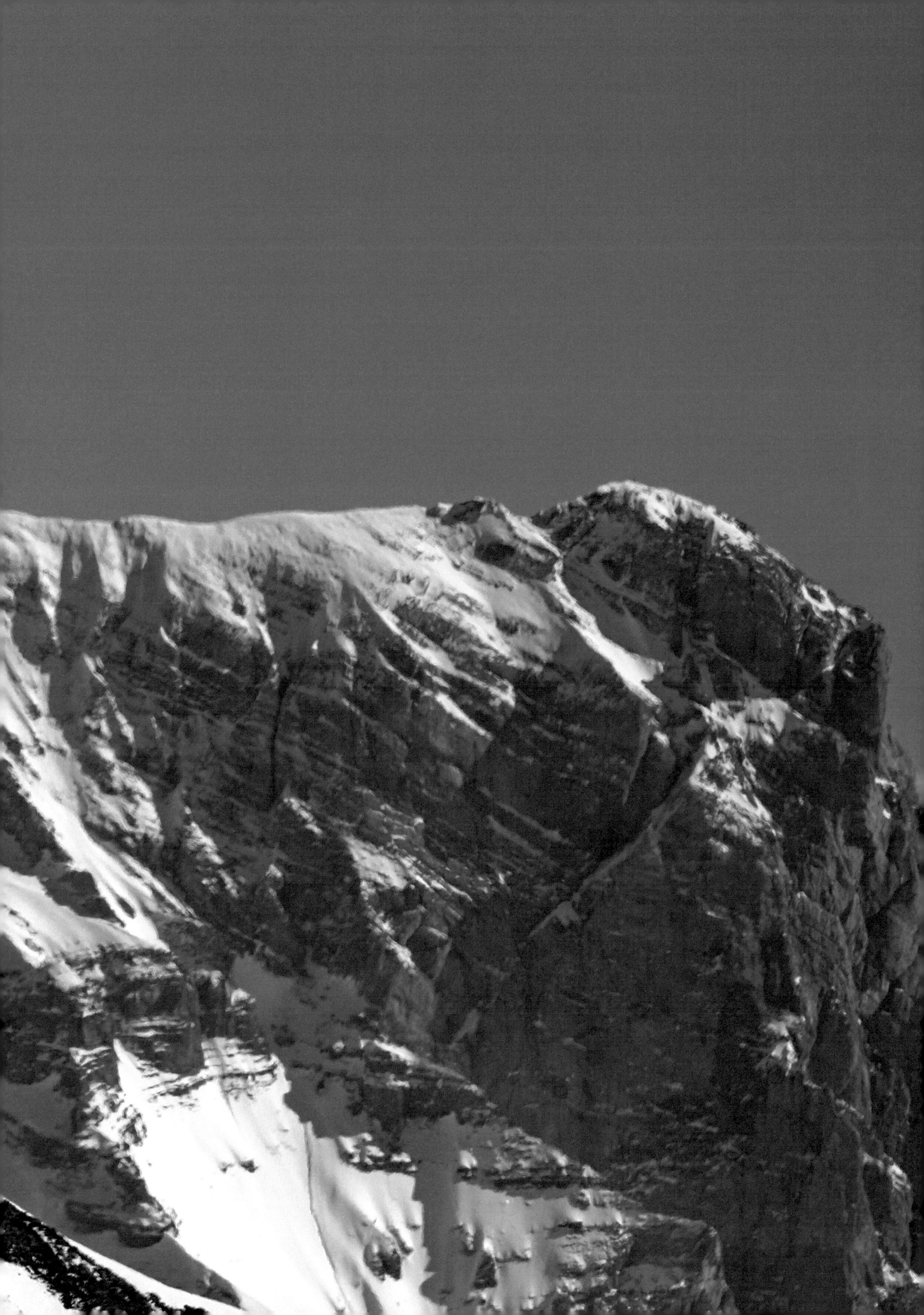

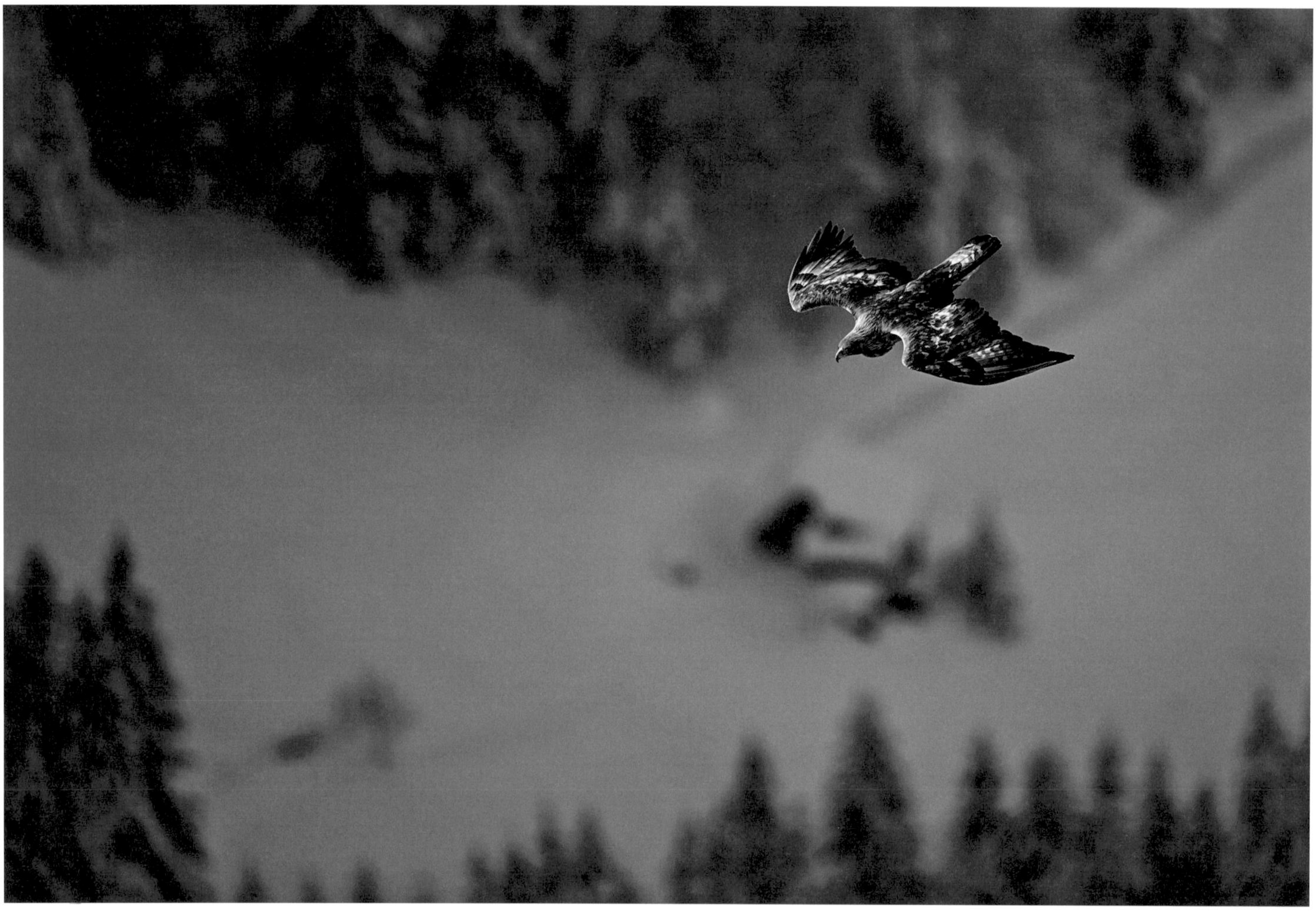

THE EAGLE ON YGGDRASIL, THE TREE OF LIFE

SNORRI STURLUSON \
1179 — 1241 \ 112 \

Much is to be told of it.
An eagle sits in the limbs of the Ash,
and he has understanding
of many a thing.

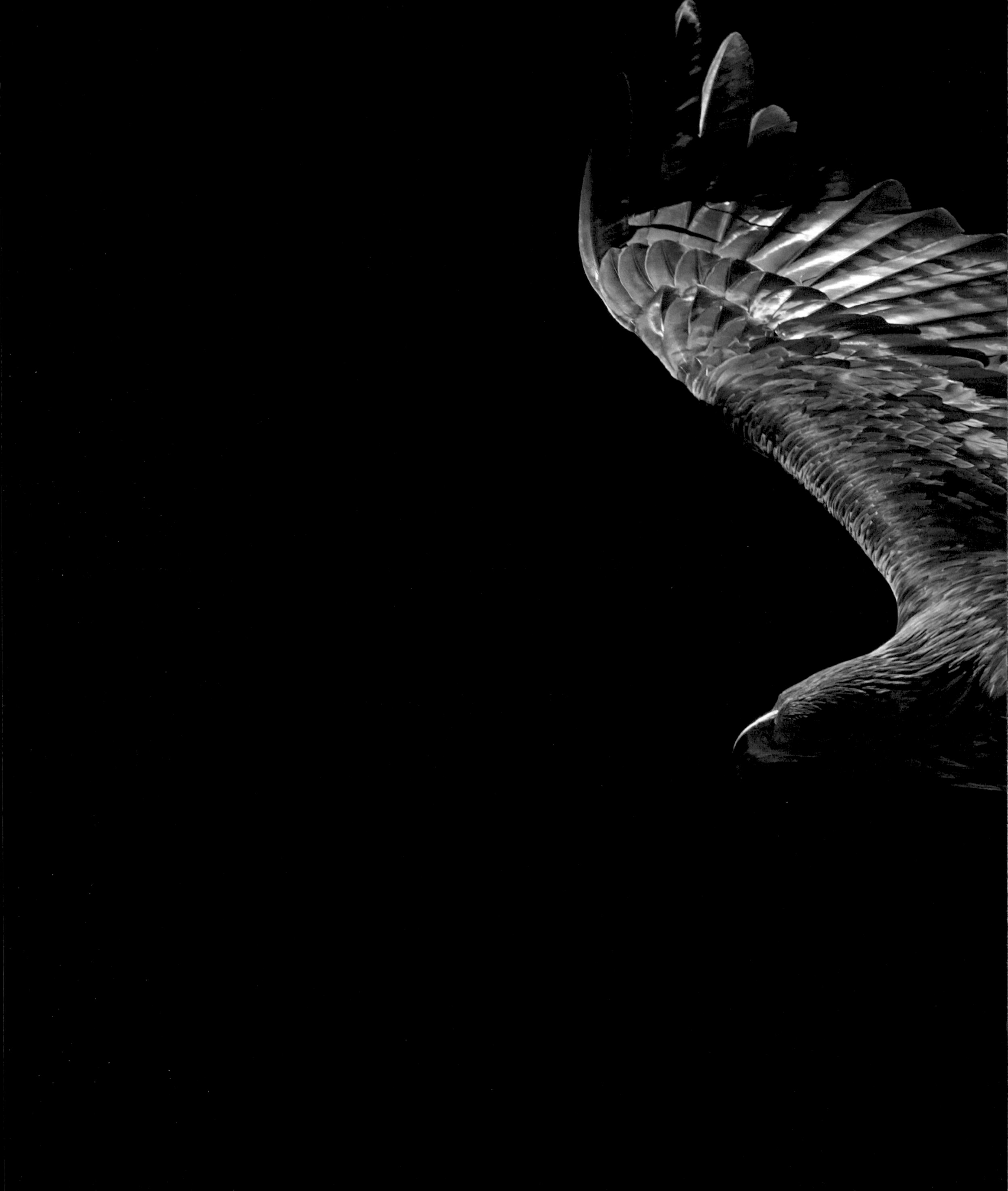

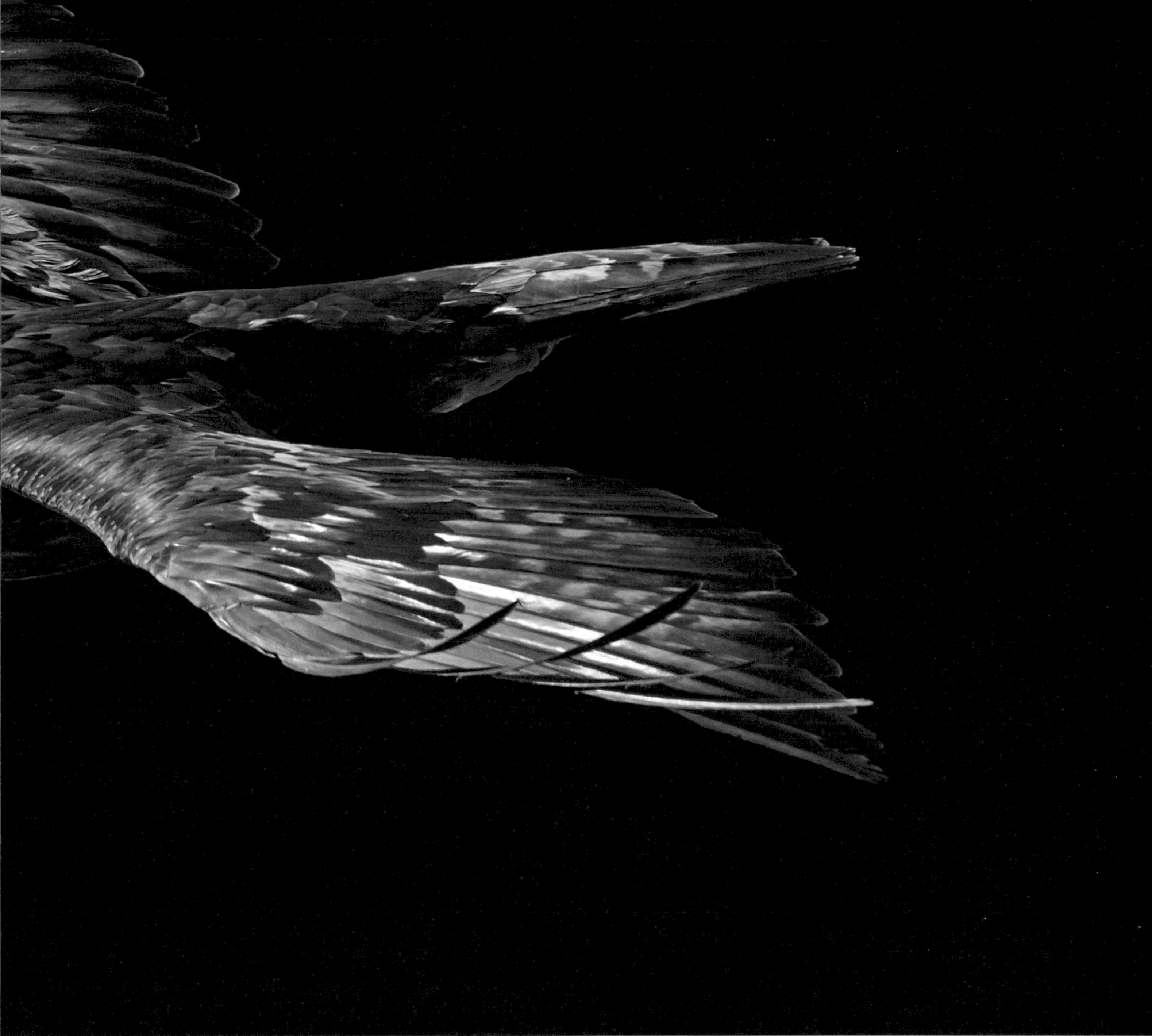

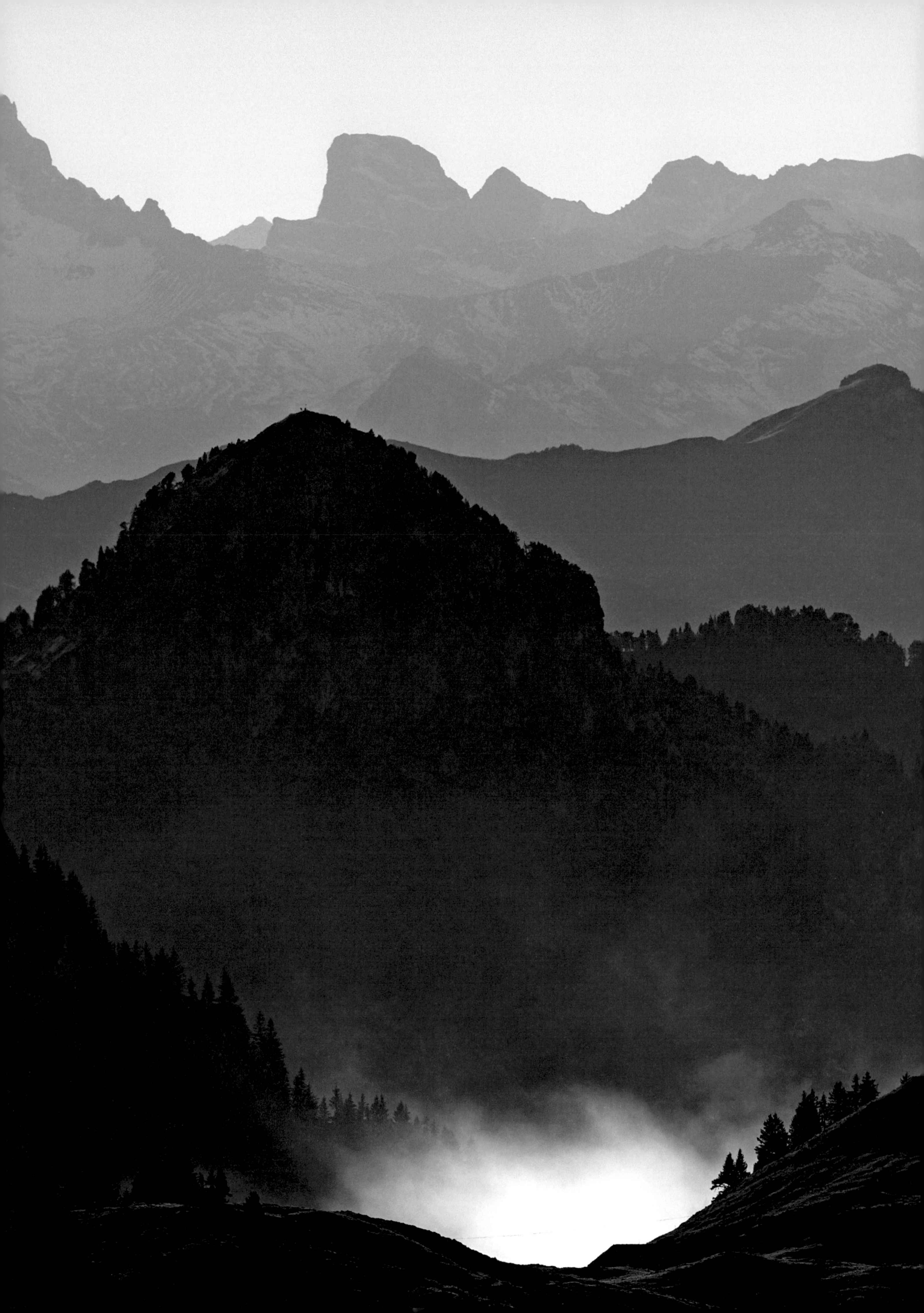

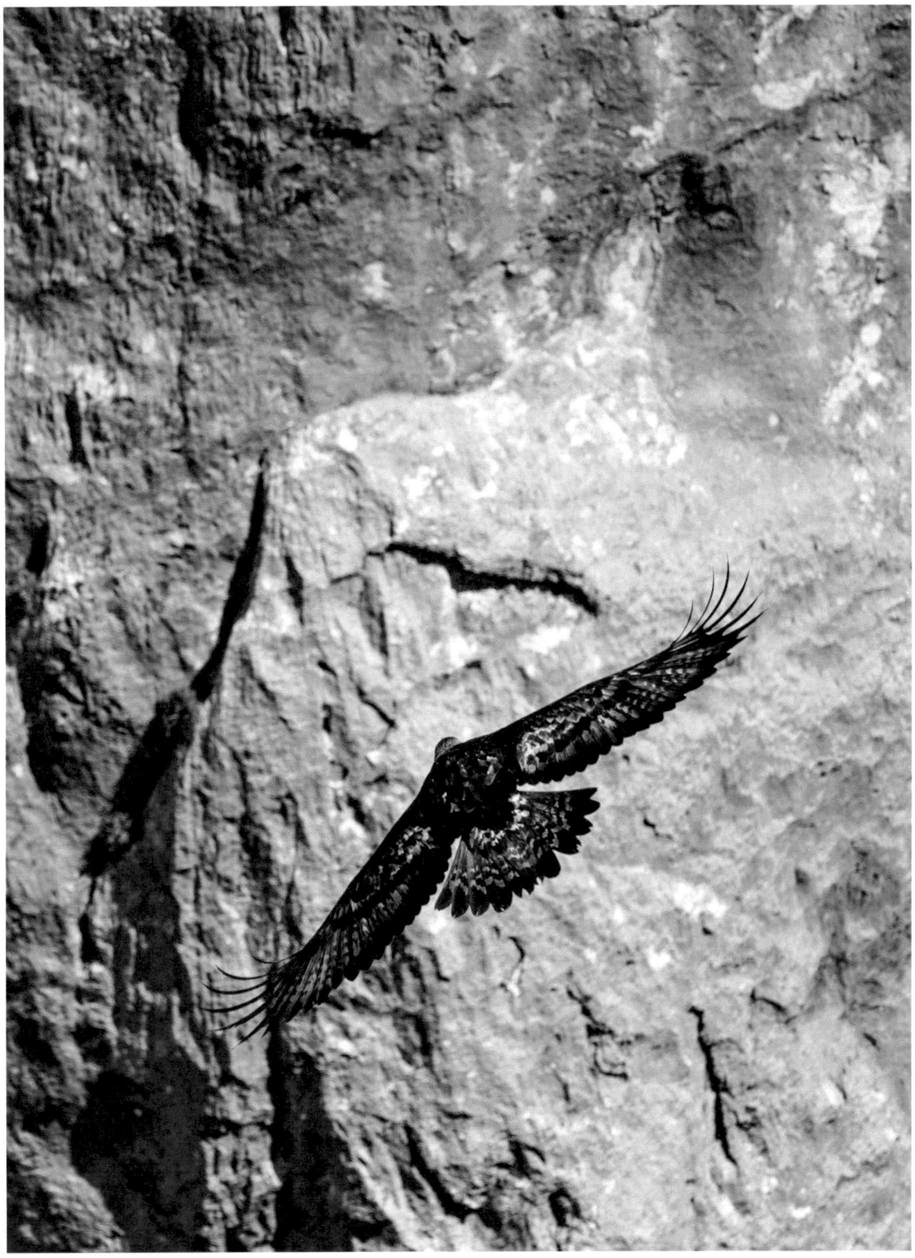

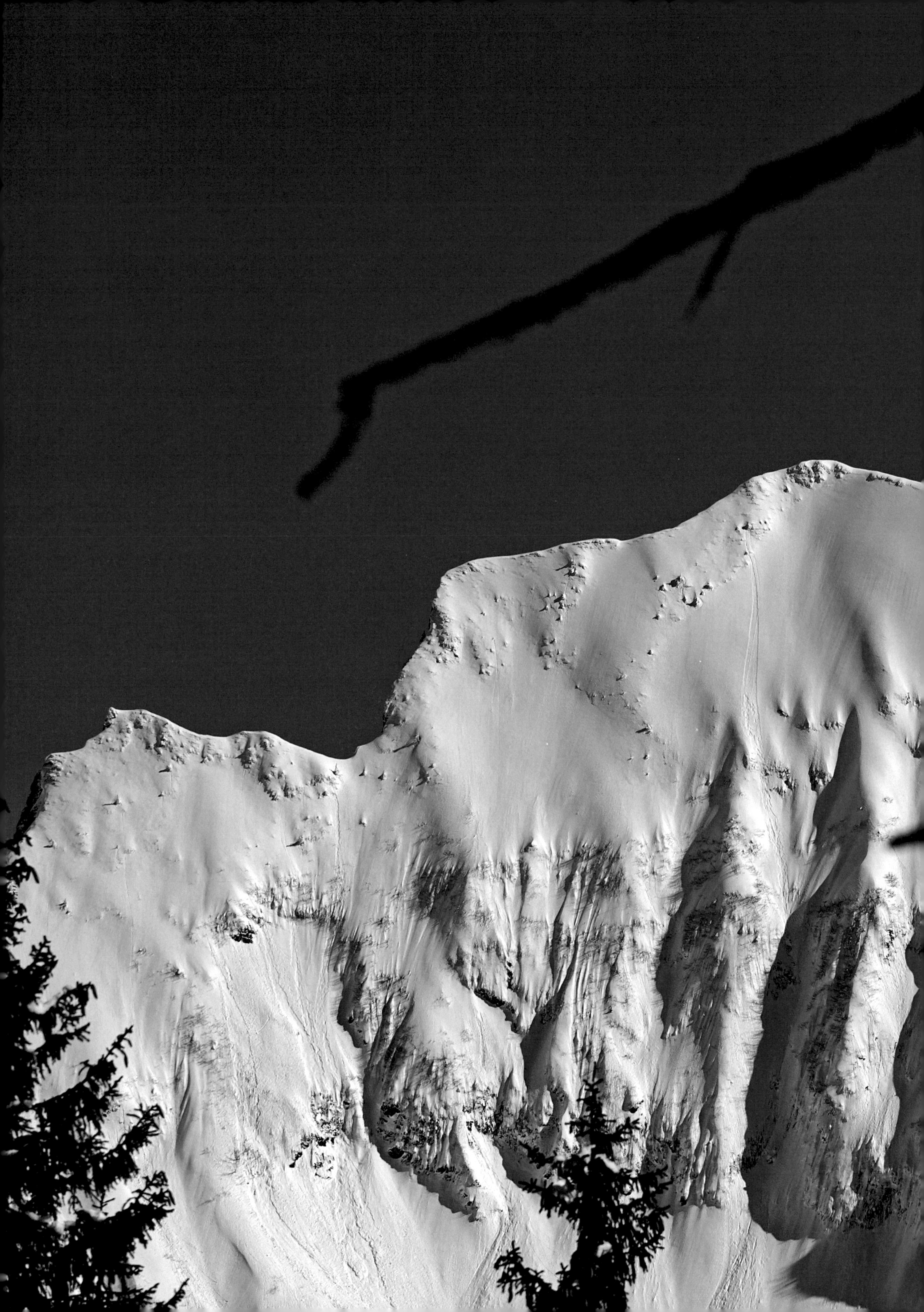

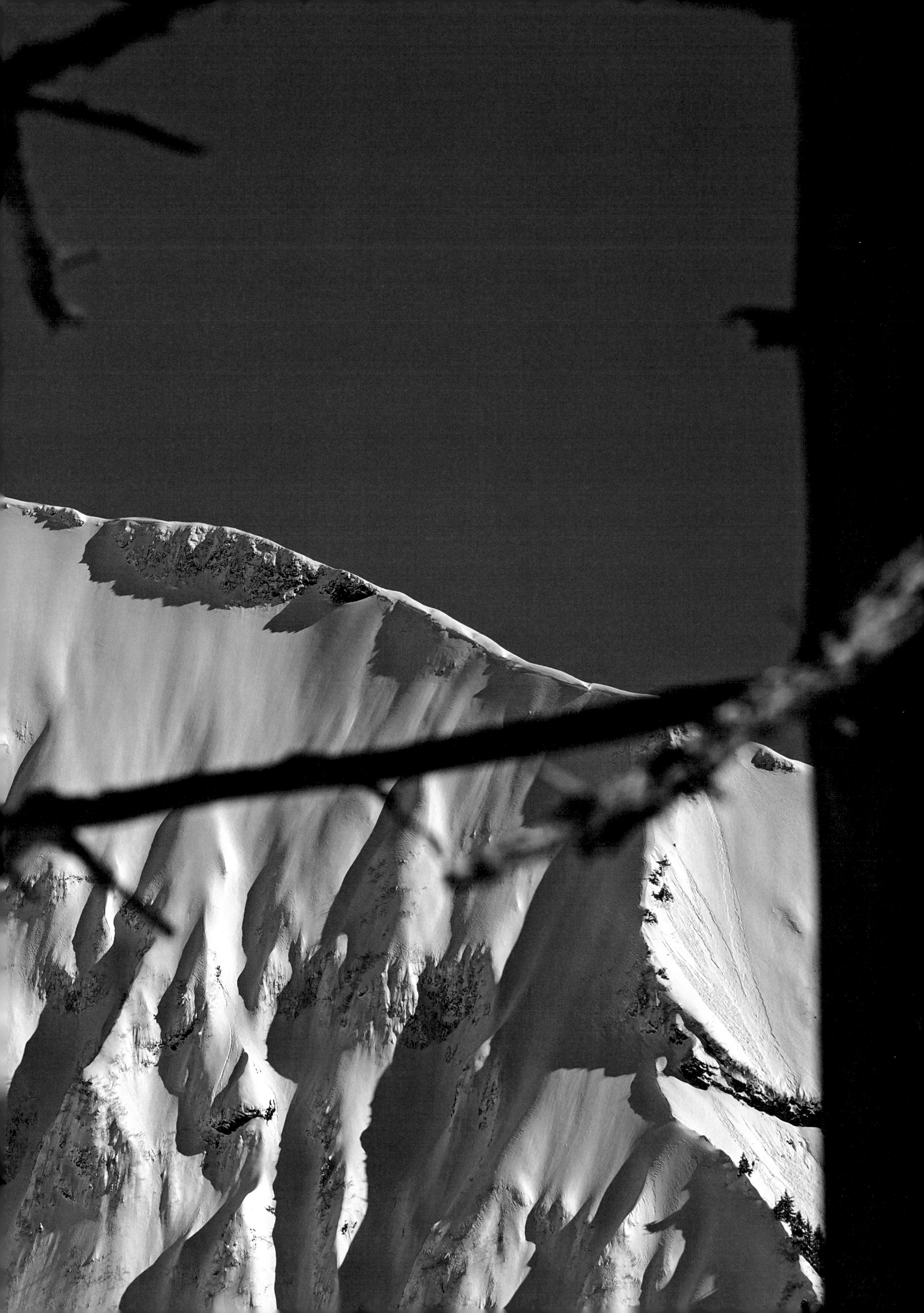

PROMETHEUS' PUNISHMENT

AESCHYLOS \
525 — 456 BC \ 113 \

This rocky chasm shall the Father split
With earthquake thunder and his burning bolt,
And he shall hide thy form, and thou shalt hang
Bolt upright, dandled in the rock's rude arms.
Nor till thou hast completed thy long term
Shalt thou come back into the light; and then
The winged hound of Zeus, the tawny eagle,
Shall violently fall upon thy flesh
And rend it as 'twere rags; and every day
And all day long shall thine unbidden guest
Sit at thy table, feasting on thy liver
Till he hath gnawn it black. Look for no term
To such an agony till there stand forth
Among the Gods one who shall take upon him
Thy sufferings and consent to enter hell
Far from the light of Sun, yea, the deep pit
And mirk of Tartarus, for thee.

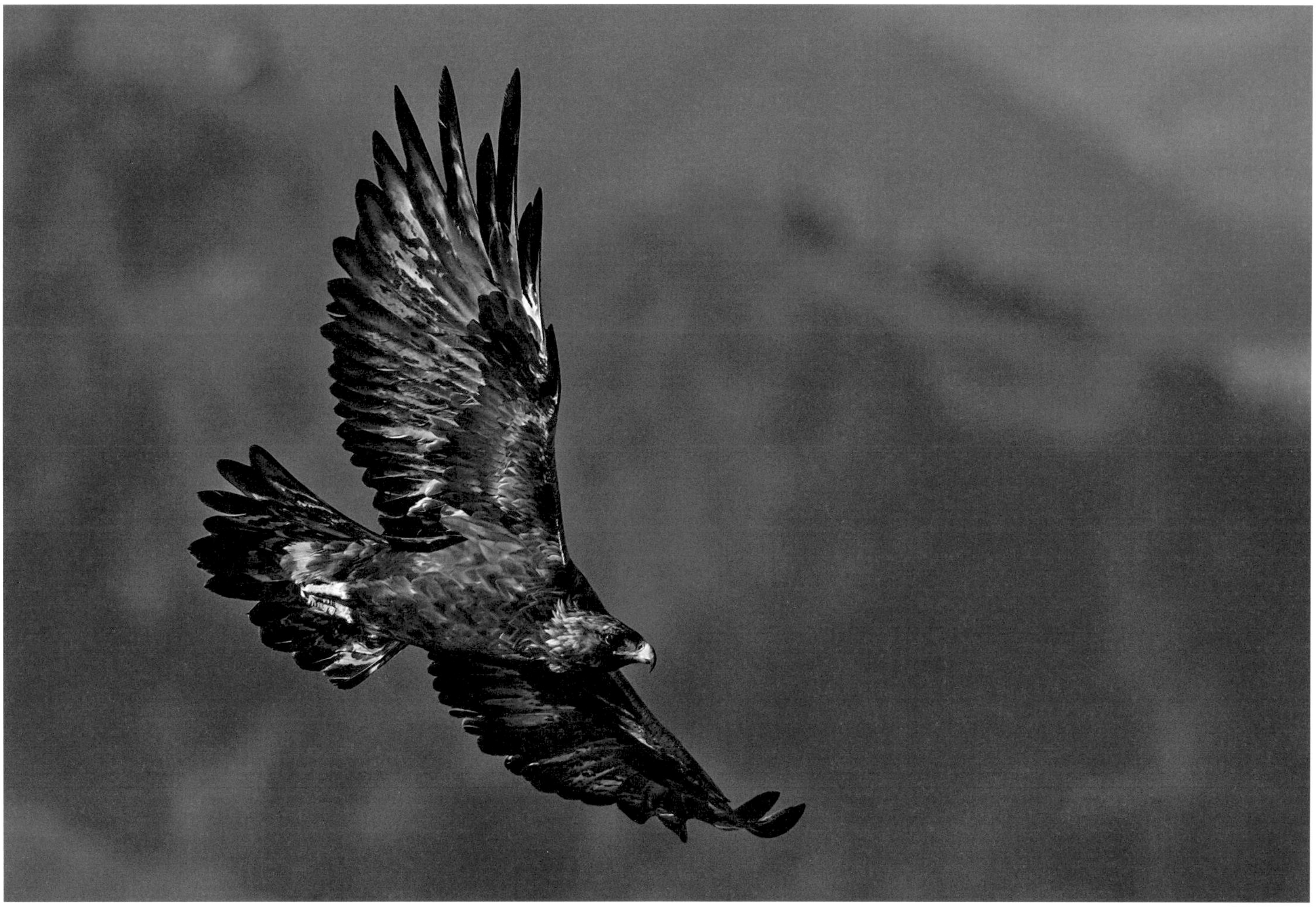

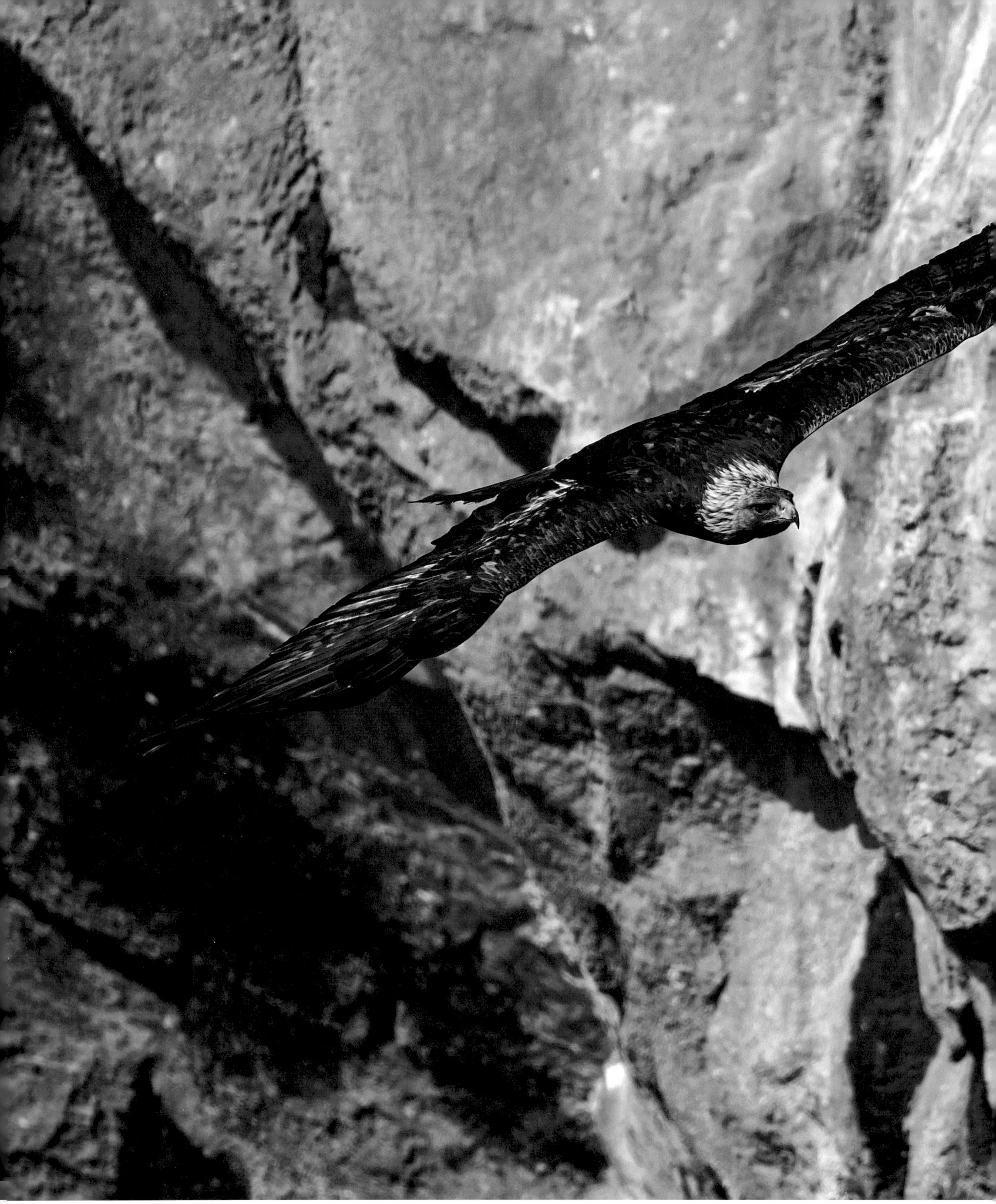

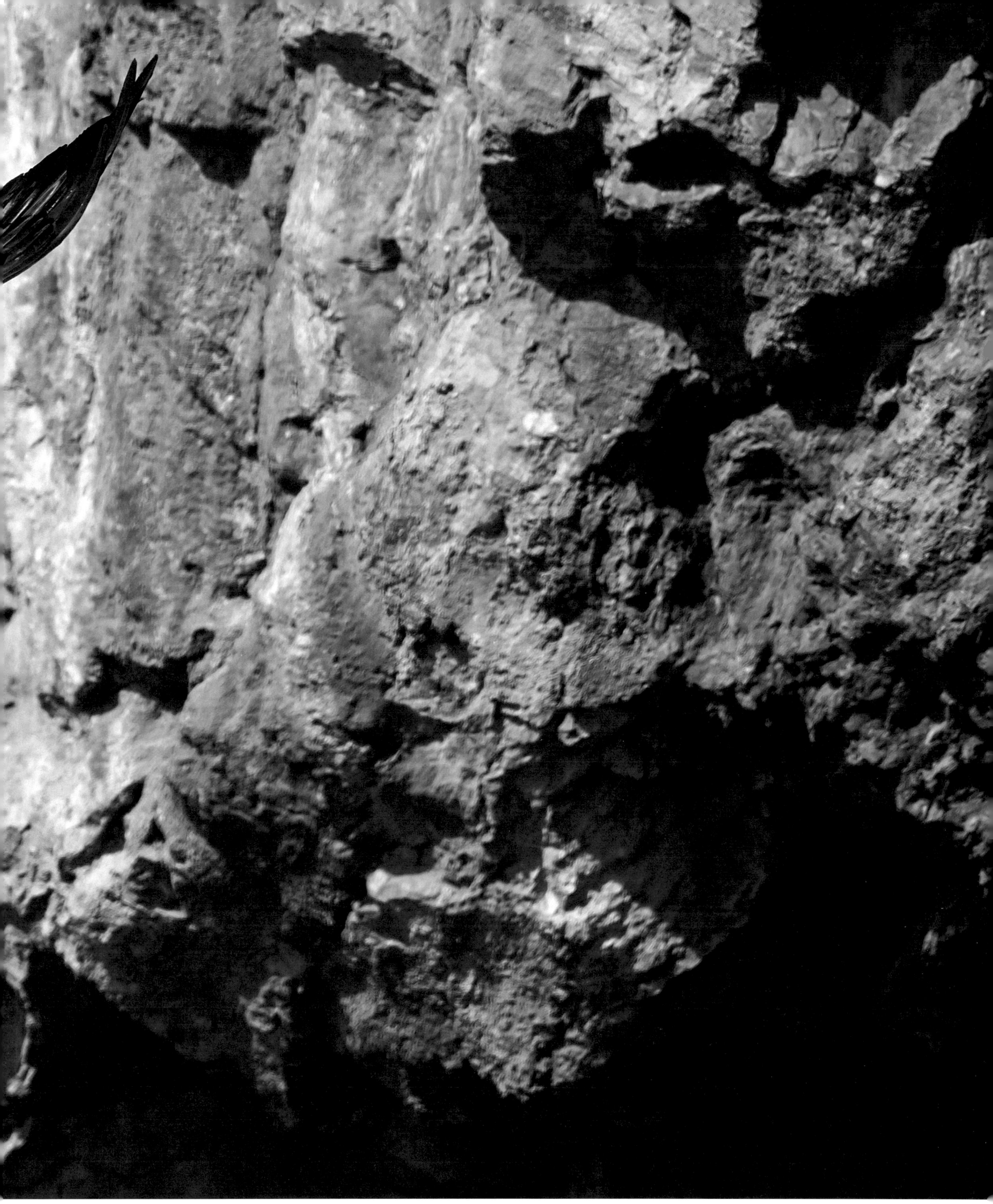

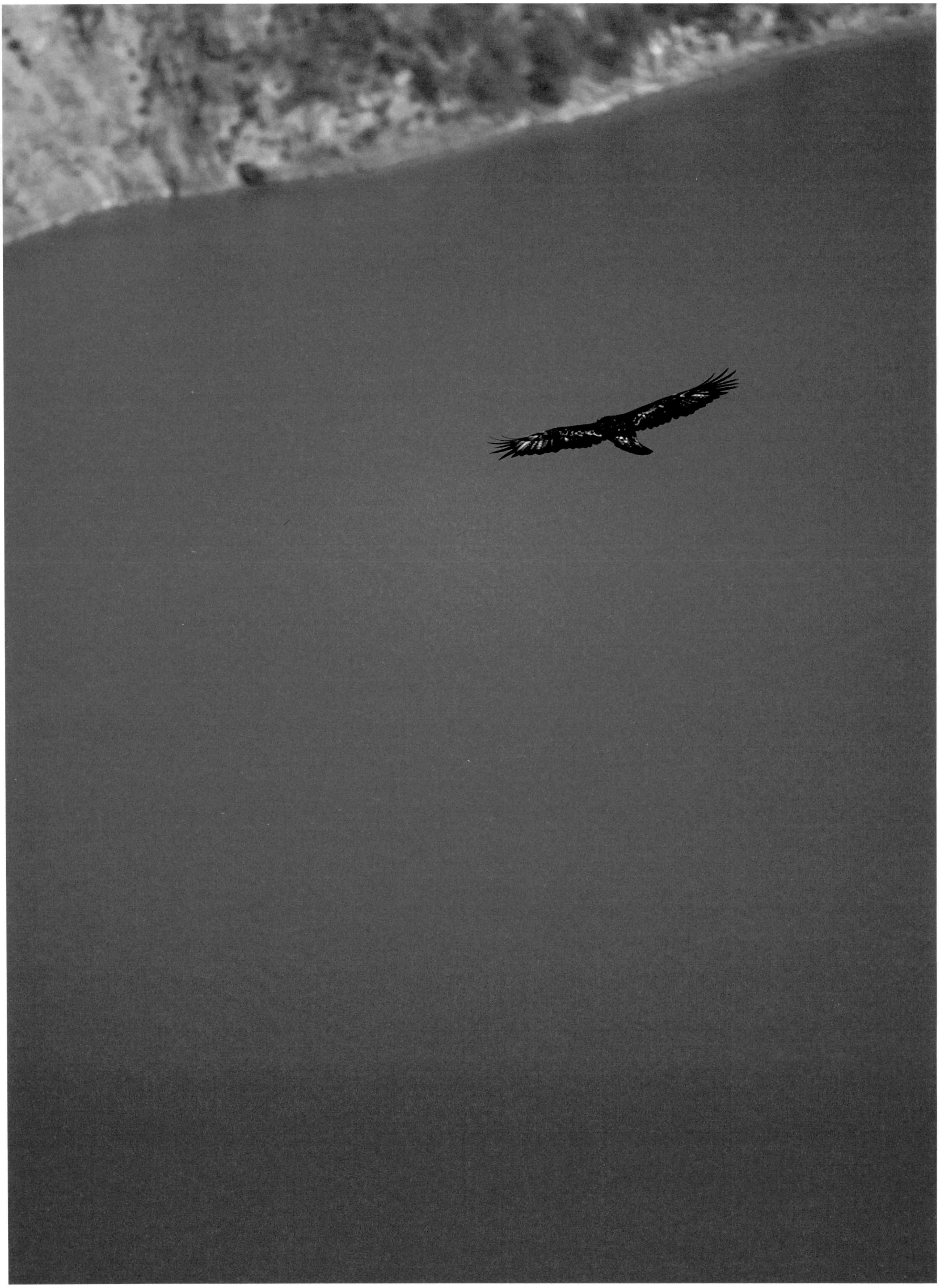

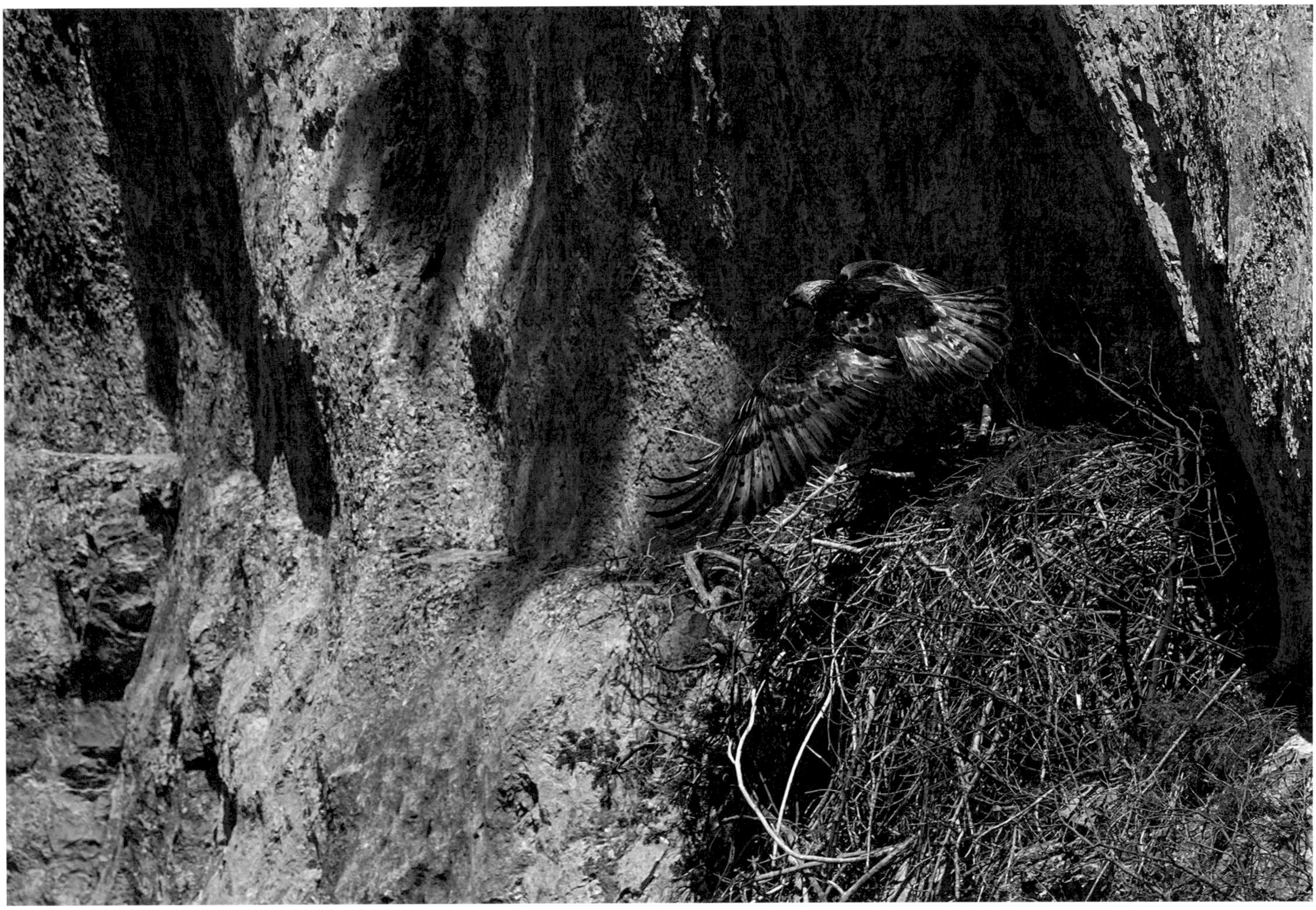

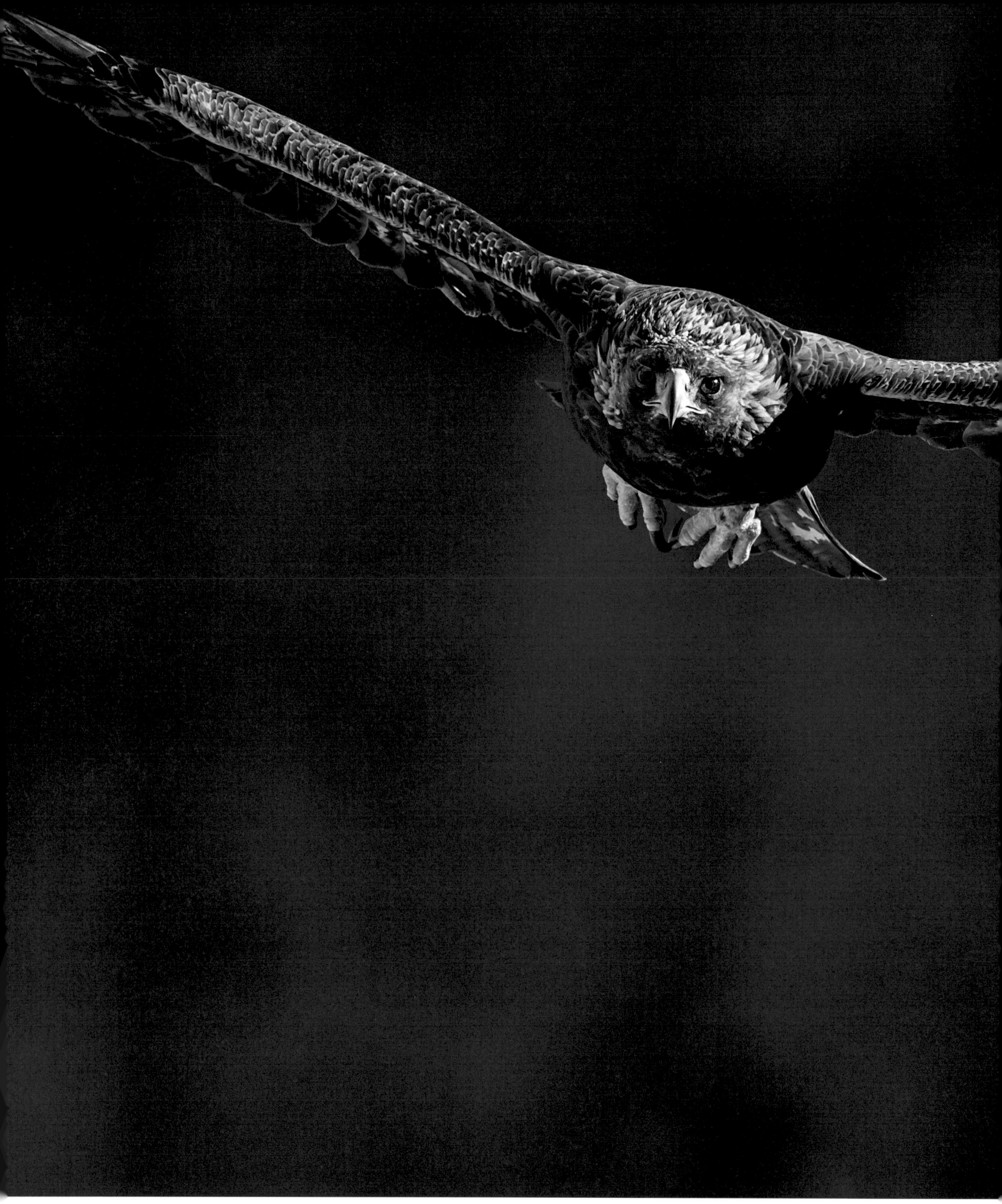

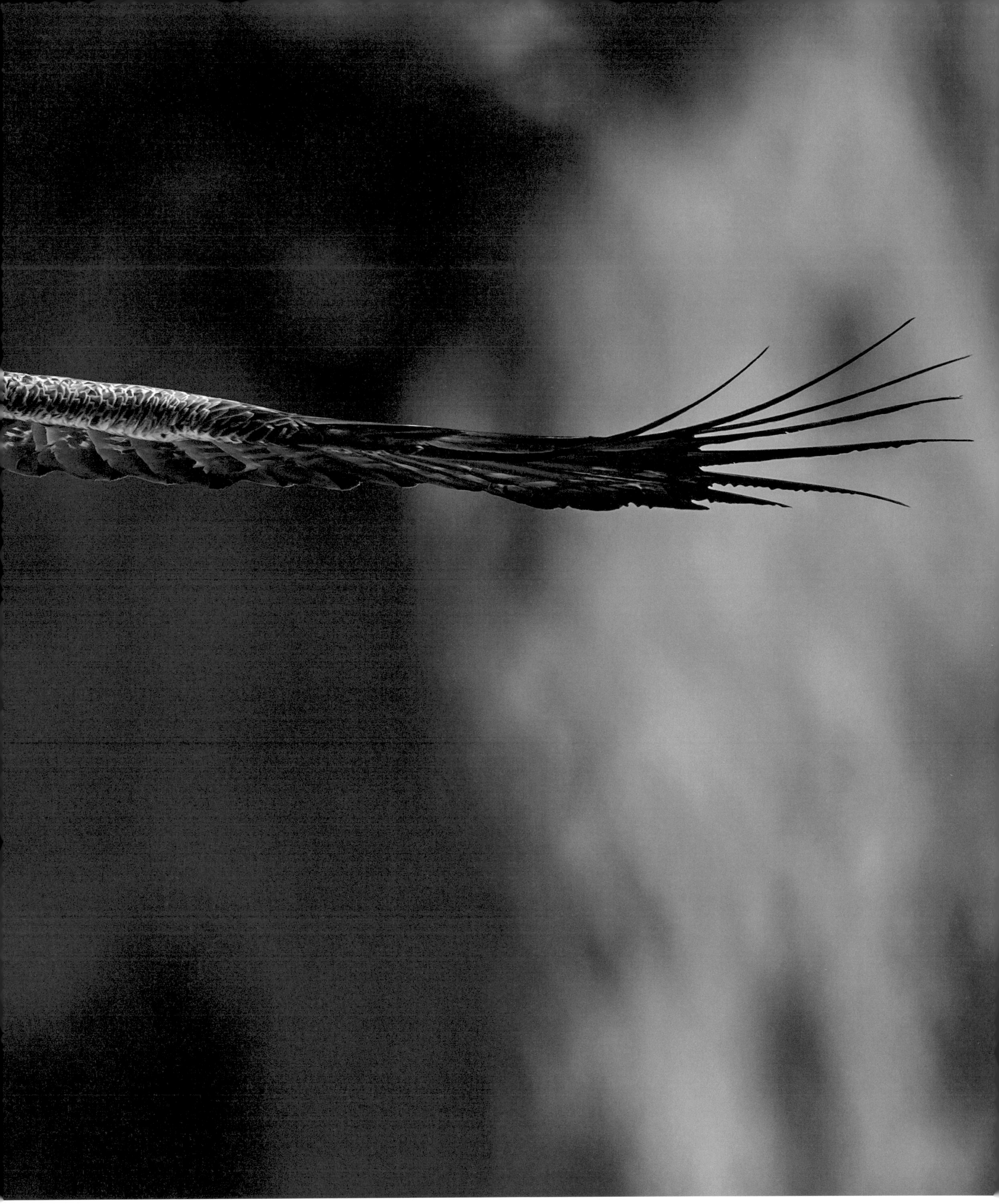

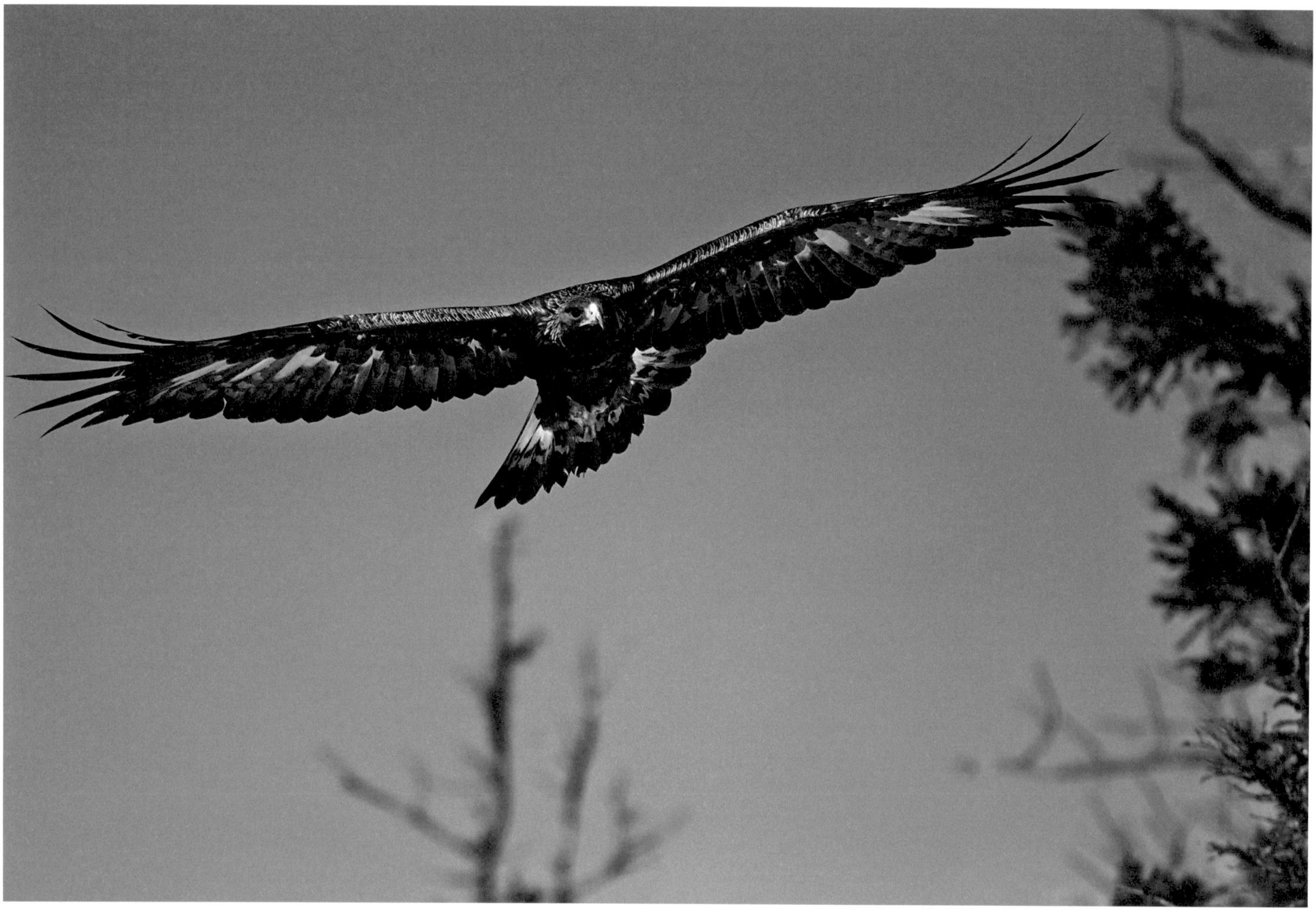

LORD OF THE BOUNDLESS REALM OF AIR

JAMES GATES PERCIVAL \
1795 — 1856 \ 114 \

To the Eagle

Bird of the broad and sweeping wing!
Thy home is high in heaven,
Where wide the storms their banners fling,
And the tempest clouds are driven.
Thy throne is on the mountain top;
Thy fields the boundless air;
And hoary peaks, that proudly prop
The skies, thy dwellings are.

Thou sittest, like a thing of light,
Amid the noontide blaze;
The midway sun is clear and bright –
It cannot dim thy gaze.
Thy pinions, to the rushing blast
O'er the bursting billow spread,
Where the vessel plunges, hurry past,
Like an angel of the dead.

Thou art perched aloft on the beetling crag,
And the waves are white below,
And on, with a haste that cannot lag,
They rush in an endless flow.
Again thou hast plumed thy wing for flight
To lands beyond the sea,
And away, like a spirit wreathed in light,
Thou hurriest wild and free.

Thou hurriest over the myriad waves,
And thou leavest them all behind;
Thou sweepest that place of unknown graves,
Fleet as the tempest wind.
When the night-storm gathers dim and dark,
With a shrill and a boding scream,
Thou rushest by the foundering bark,
Quick'as a passing dream.

Lord of the boundless realm of air!
In thy imperial name
The hearts of the bold and ardent dare
The dangerous path of fame.
Beneath the shade of thy golden wings,
The Roman legions bore,
From the river of Egypt's cloudy springs,
Their pride, to the polar shore.

For thee they fought, for thee they fell,
And their oath was on thee laid:
To thee the clarions raised their swell,
And the dying warrior prayed.
Thou wert, through an age of death and fears,
The image of pride and power,
Till the gathered rage of a thousand years
Burst forth in one awful hour.

And then, a deluge of wrath it came,
And the nations shook with dread;
And it swept the earth, till its fields were flame,
And piled with the mingled dead.
Kings were rolled in the wasteful flood,
With the low and crouching slave;
And together lay, in a shroud of blood,
The coward and the brave.

And where was then thy fearless flight?
"O'er the dark mysterious sea,
To the lands that caught the setting light,
The cradle of liberty.
There, on the silent and lonely shore,
For ages I watched alone,
And the world, in its darkness, asked no more,
Where the glorious bird had flown.

But there came a bold and hardy few,
And they breasted the unknown wave;
I caught afar the wandering crew,
And I knew they were high and brave.
I wheeled around the welcome bark,
As it sought the desolate shore,
And up to heaven, like a joyous lark,
My quivering pinions bore.

And now that bold and hardy few
Are a nation wide and strong,
And danger and doubt I have led them through,
And they worship me in song;
And over their bright and glancing arms,
On field and lake and sea,
With an eye that fires, and a spell that charms,
I guide them to victory."

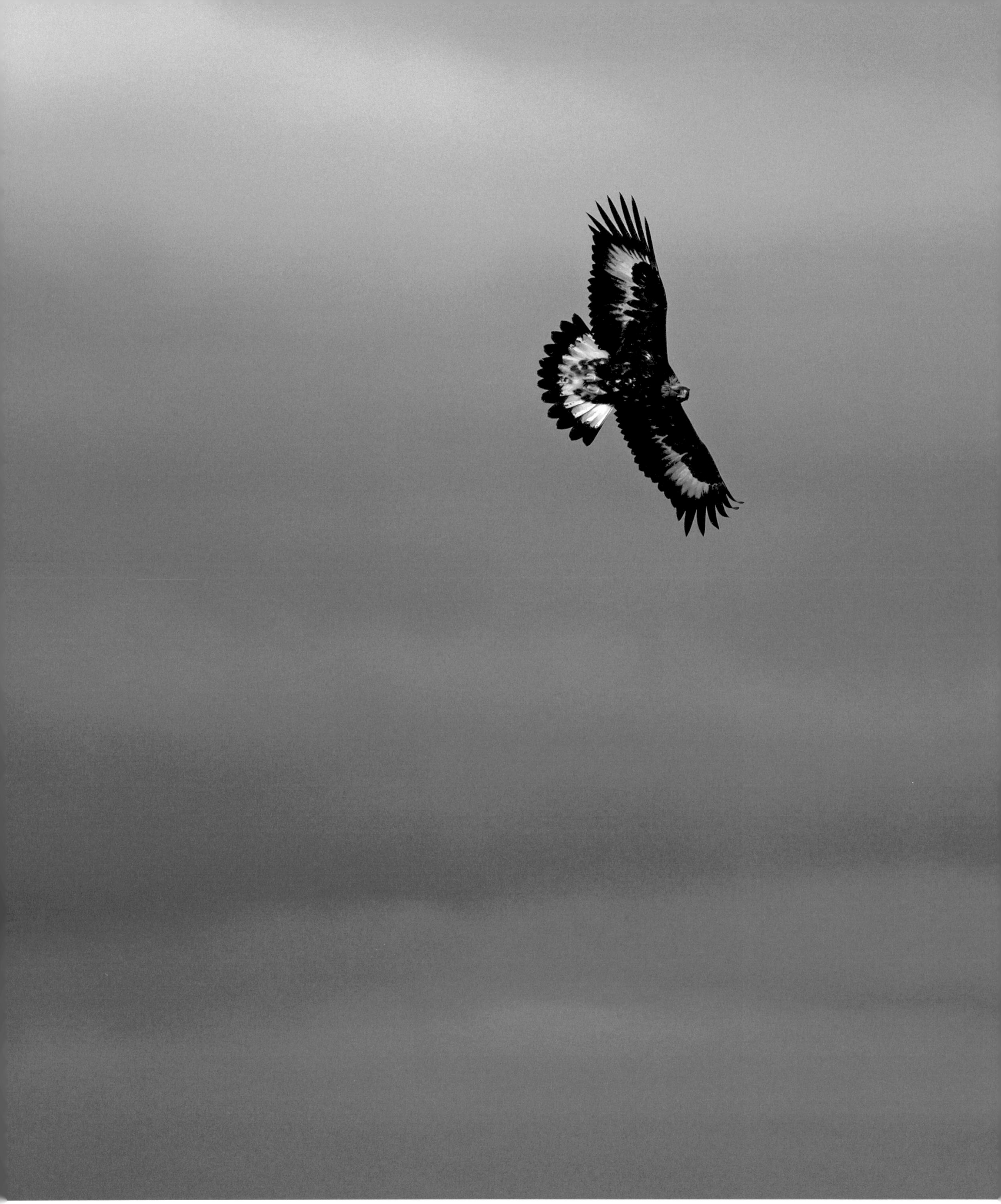

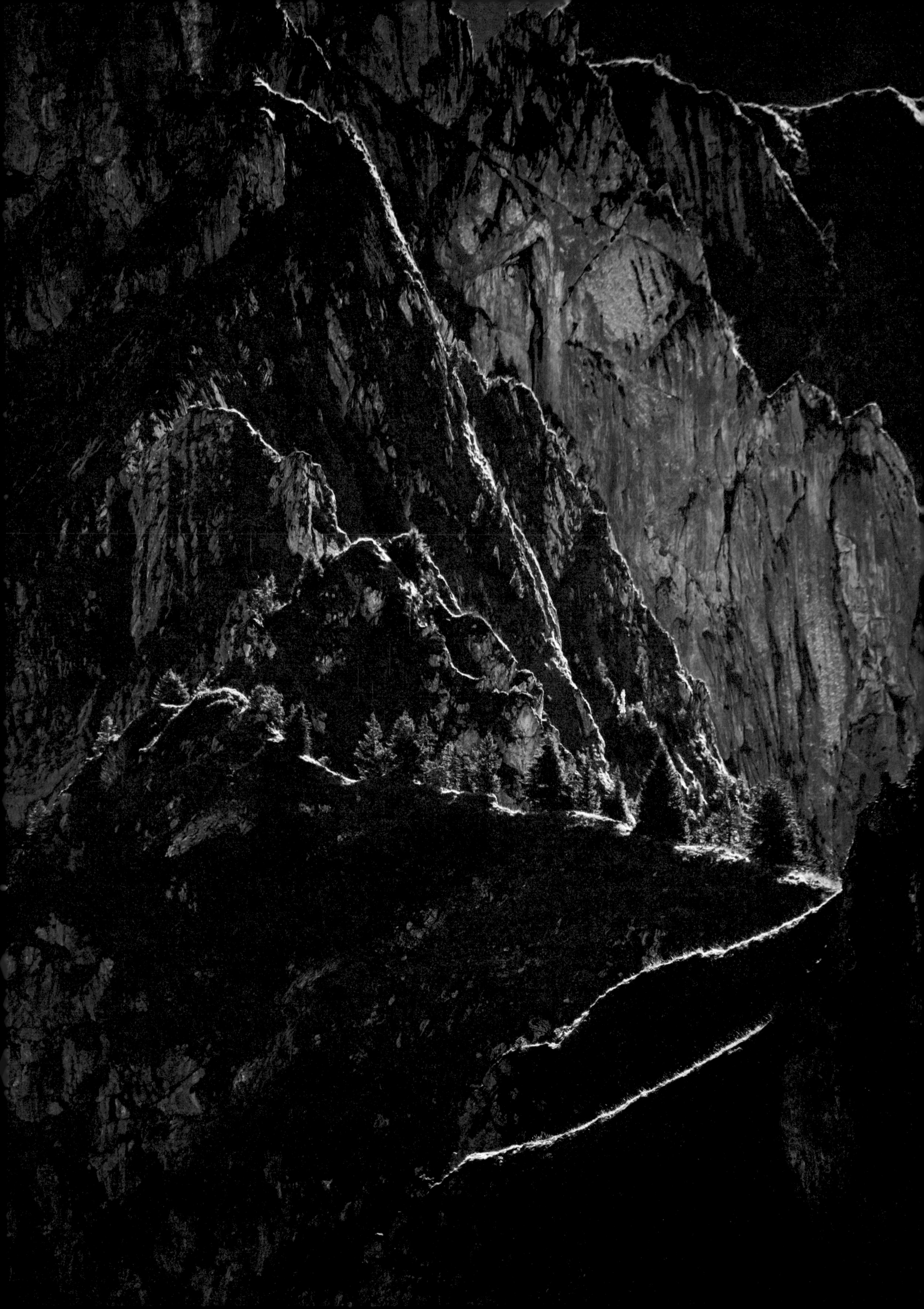

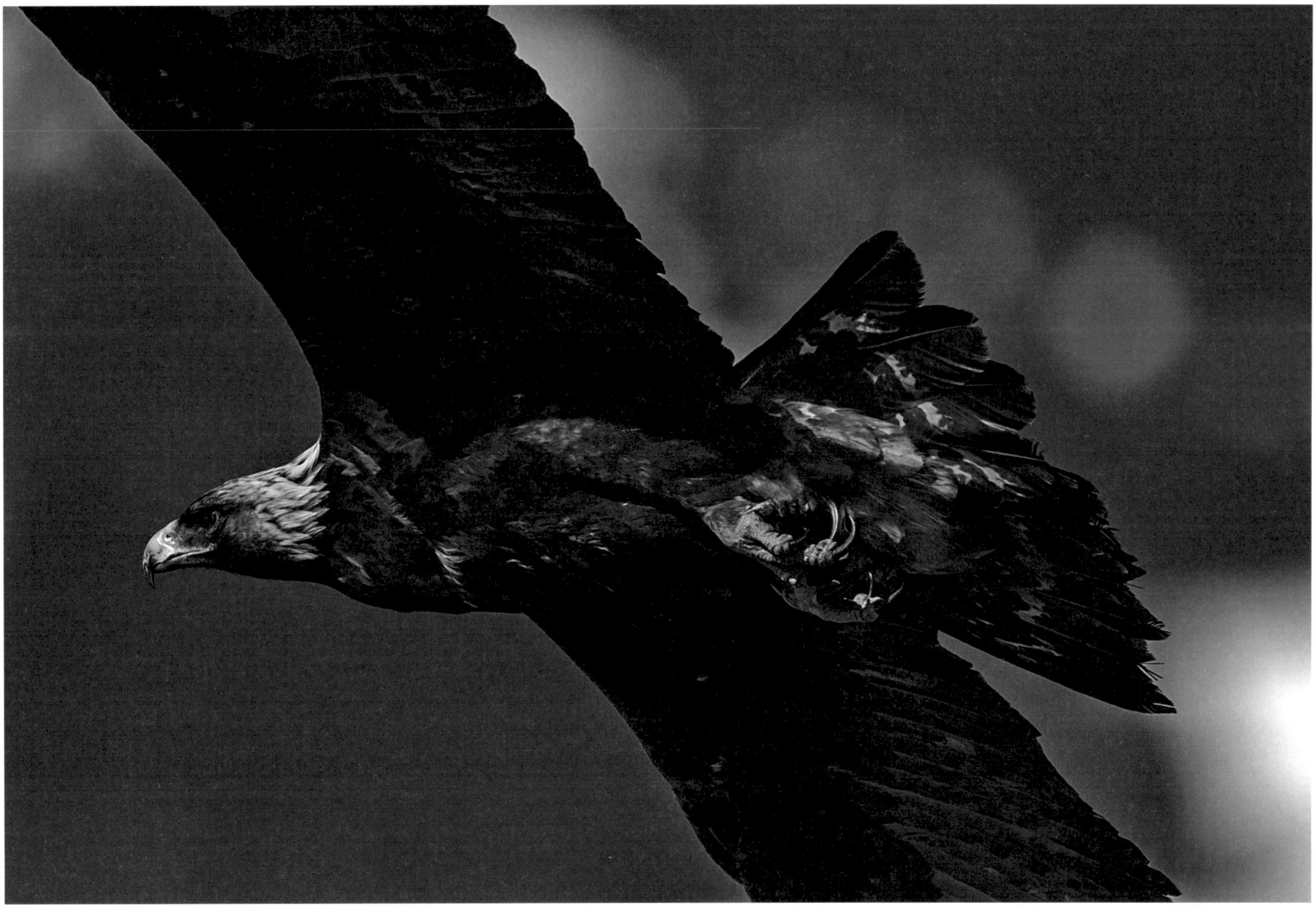

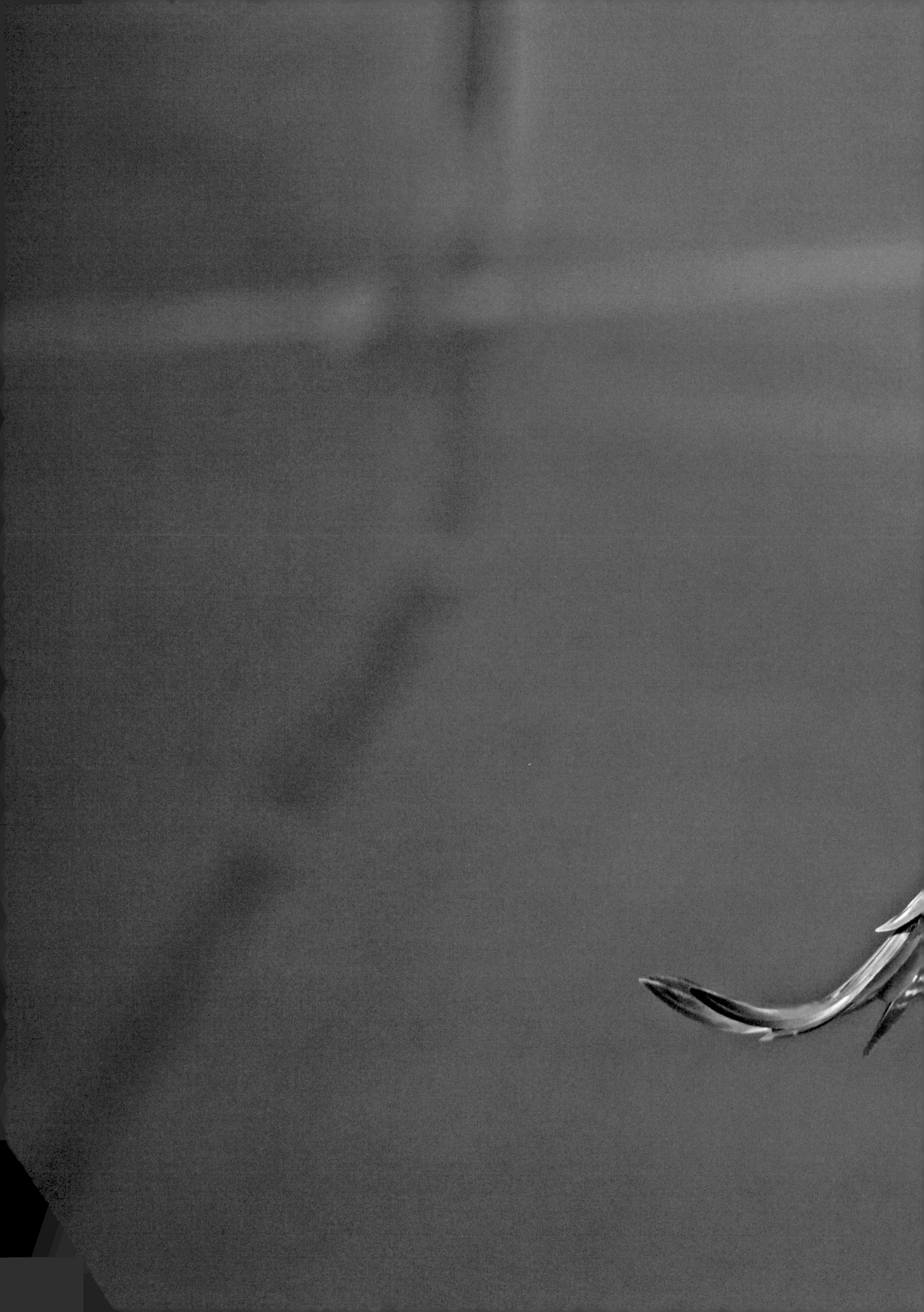

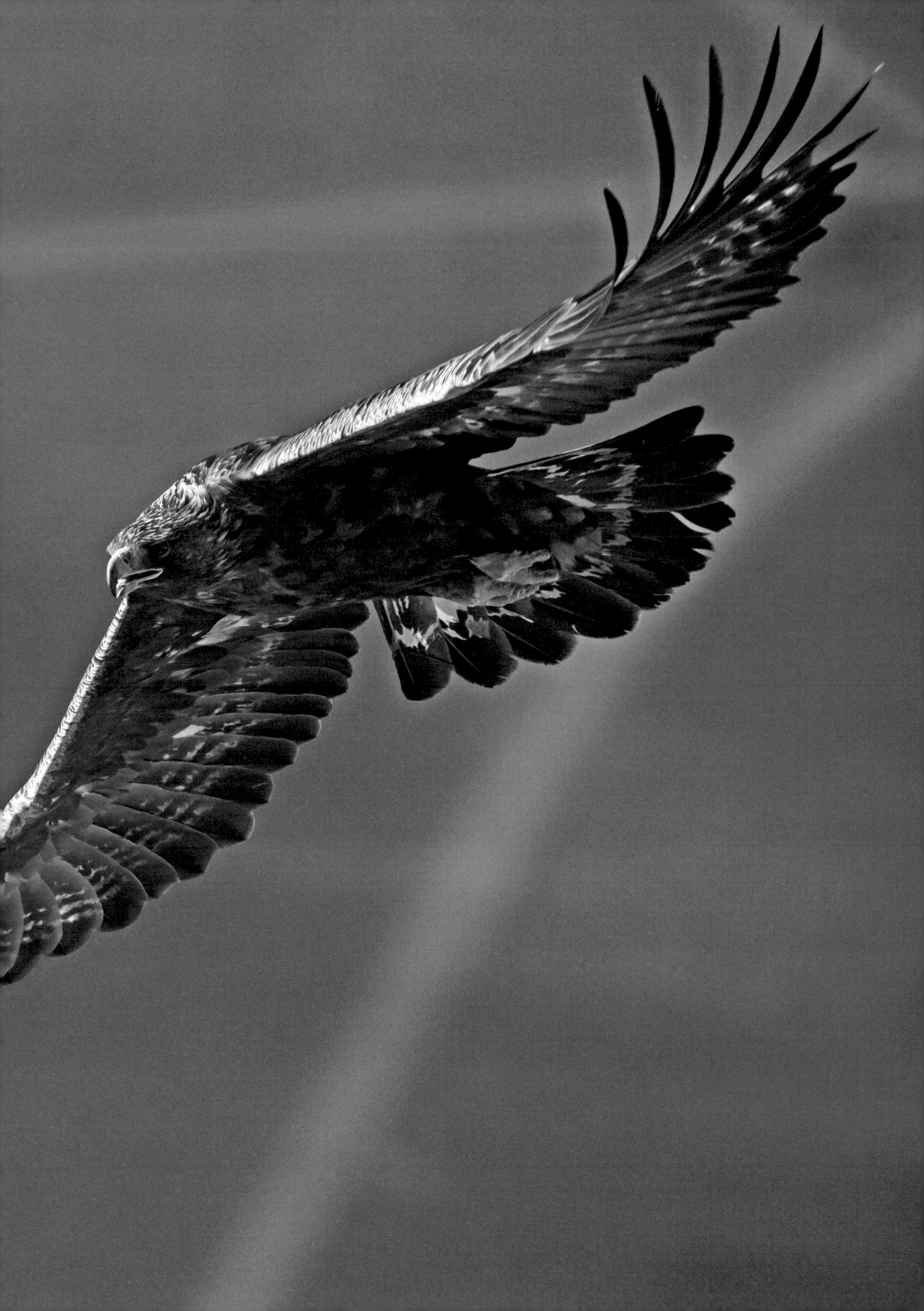

THE EAGLE AND THE SNAKE

PERCY BYSSHE SHELLEY \ 1792 — 1822 \ 115 \

The Revolt of Islam,
Canto first, VIII-XIV

A course precipitous, of dizzy speed,
Suspending thought and breath; a monstrous sight!
For in the air do I behold indeed
An Eagle and a Serpent wreathed in fight: –
And now relaxing its impetuous flight,
Before the aërial rock on which I stood,
The Eagle, hovering, wheeled to left and right,
And hung with lingering wings over the flood,
And startled with its yells the wide air's solitude.

A shaft of light upon its wings descended,
And every golden feather gleamed therein –
Feather and scale inextricably blended.
The Serpent's mailed and many-colored skin
Shone thro' the plumes its coils were twined within
By many a swollen and knotted fold, and high
And far, the neck receding lithe and thin,
Sustained a crested head, which warily
Shifted and glanced before the Eagle's stedfast eye.

Around, around, in ceaseless circles wheeling
With clang of wings and scream, the Eagle sailed
Incessantly – sometimes on high concealing
Its lessening orbs, sometimes as if it failed,
Drooped thro' the air; and still it shrieked and wailed,
And casting back its eager head, with beak
And talon unremittingly assailed
The wreathed Serpent, who did ever seek
Upon his enemy's heart a mortal wound to wreak.

What life what power was kindled and arose
Within the sphere of that appalling fray!
For, from the encounter of those wond'rous foes,
A vapor like the sea's suspended spray
Hung gathered; in the void air, far away,
Floated the shattered plumes; bright scales did leap,
Where'er the Eagle's talons made their way,
Like sparks into the darkness; – as they sweep,
Blood stains the snowy foam of the tumultuous deep.

Swift chances in that combat – many a check,
And many a change, a dark and wild turmoil;
Sometimes the Snake around his enemy's neck
Locked in stiff rings his adamantine coil,
Until the Eagle, faint with pain and toil,
Remitted his strong flight, and near the sea
Languidly fluttered, hopeless so to foil
His adversary, who then reared on high
His red and burning crest, radiant with victory.

Then on the white edge of the bursting surge,
Where they had sunk together, would the Snake
Relax his suffocating grasp, and scourge
The wind with his wild writhings; for to break
That chain of torment, the vast bird would shake
The strength of his unconquerable wings
As in despair, and with his sinewy neck,
Dissolve in sudden shock those linked rings,
Then soar – as swift as smoke from a volcano springs.

Wile baffled wile, and strength encountered strength,
Thus long, but unprevailing: – the event
Of that portentous fight appeared at length:
Until the lamp of day was almost spent
It had endured, when lifeless, stark and rent,
Hung high that mighty Serpent, and at last
Fell to the sea, while o'er the continent
With clang of wings and scream the Eagle past,
Heavily borne away on the exhausted blast.

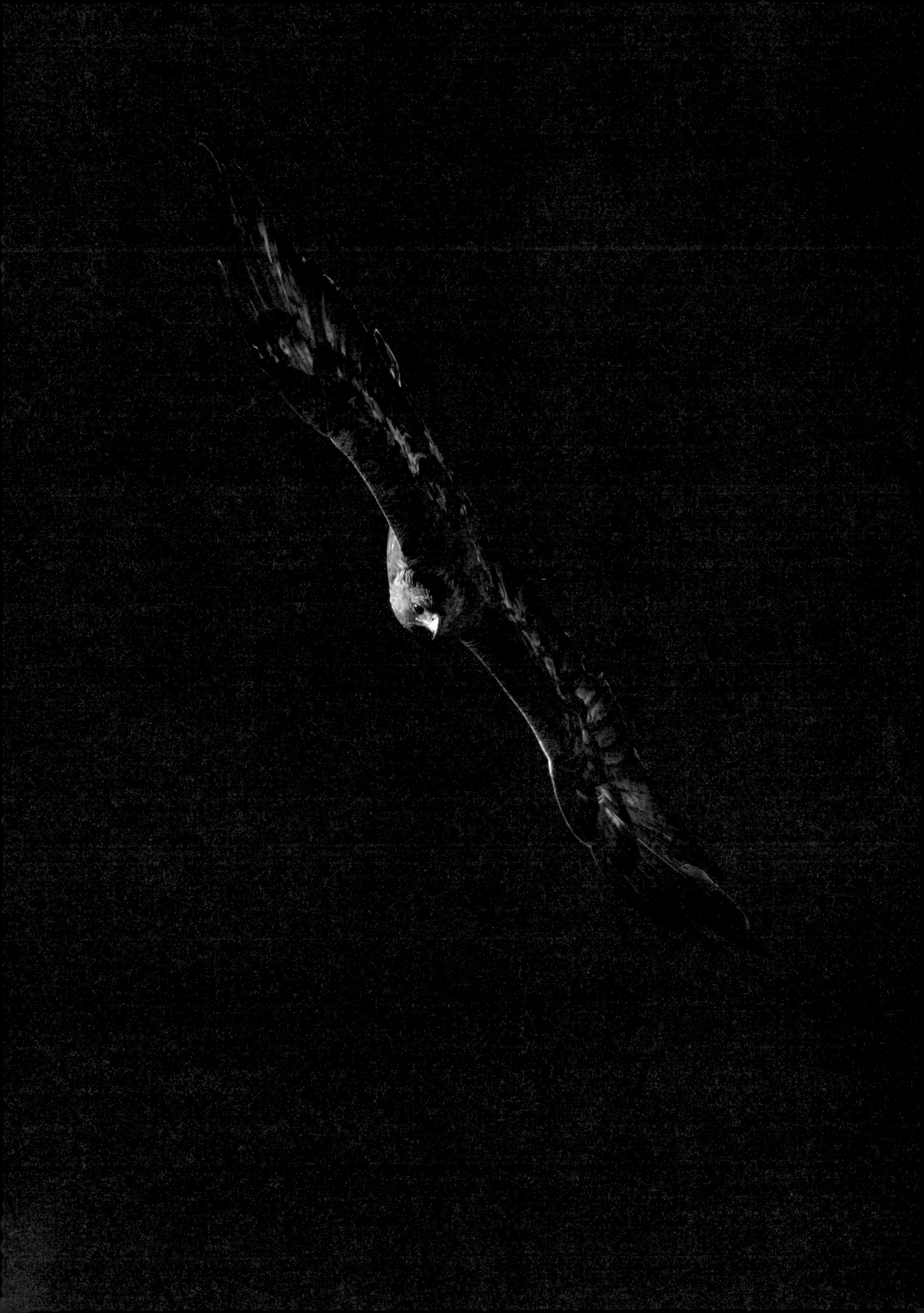

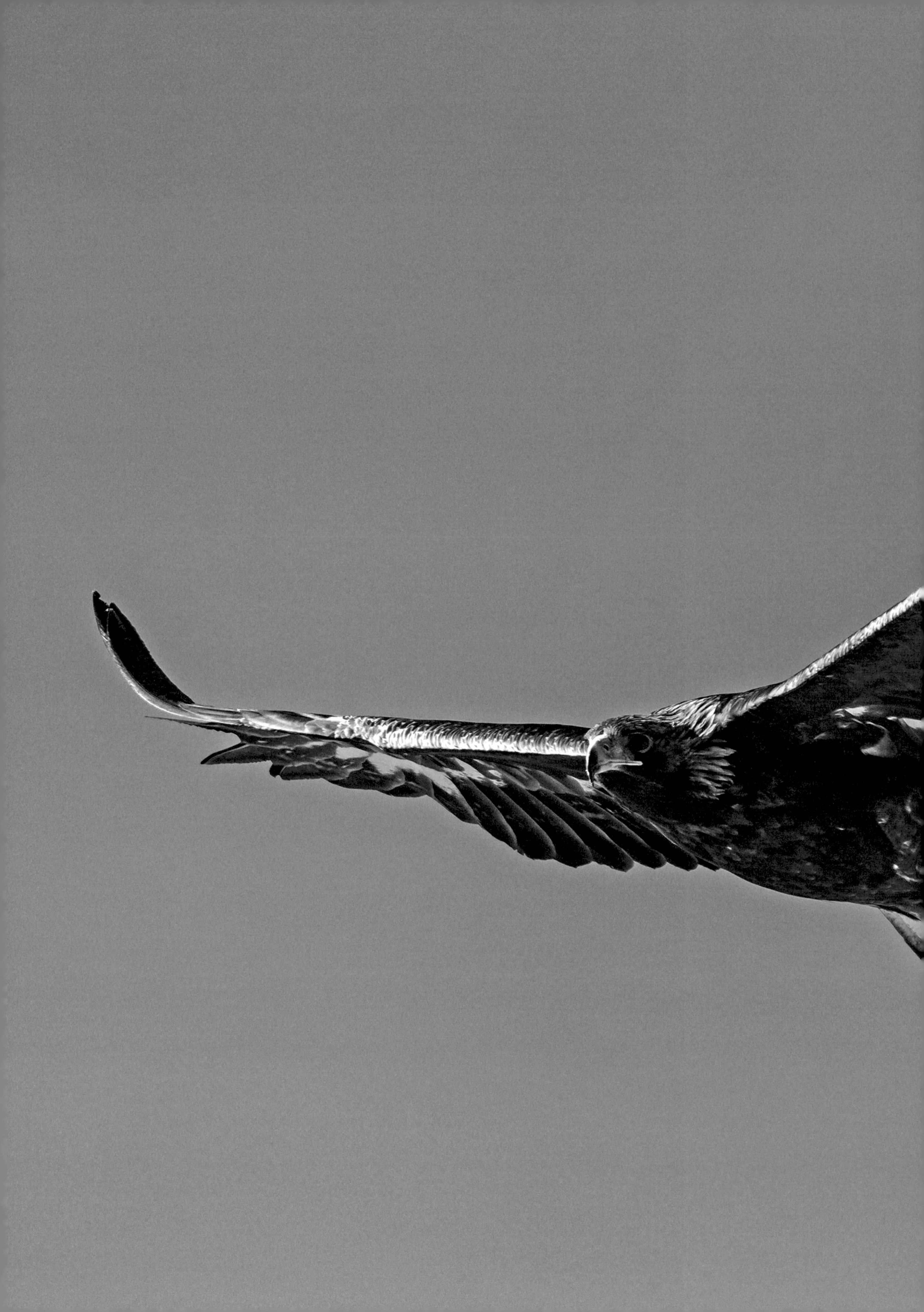

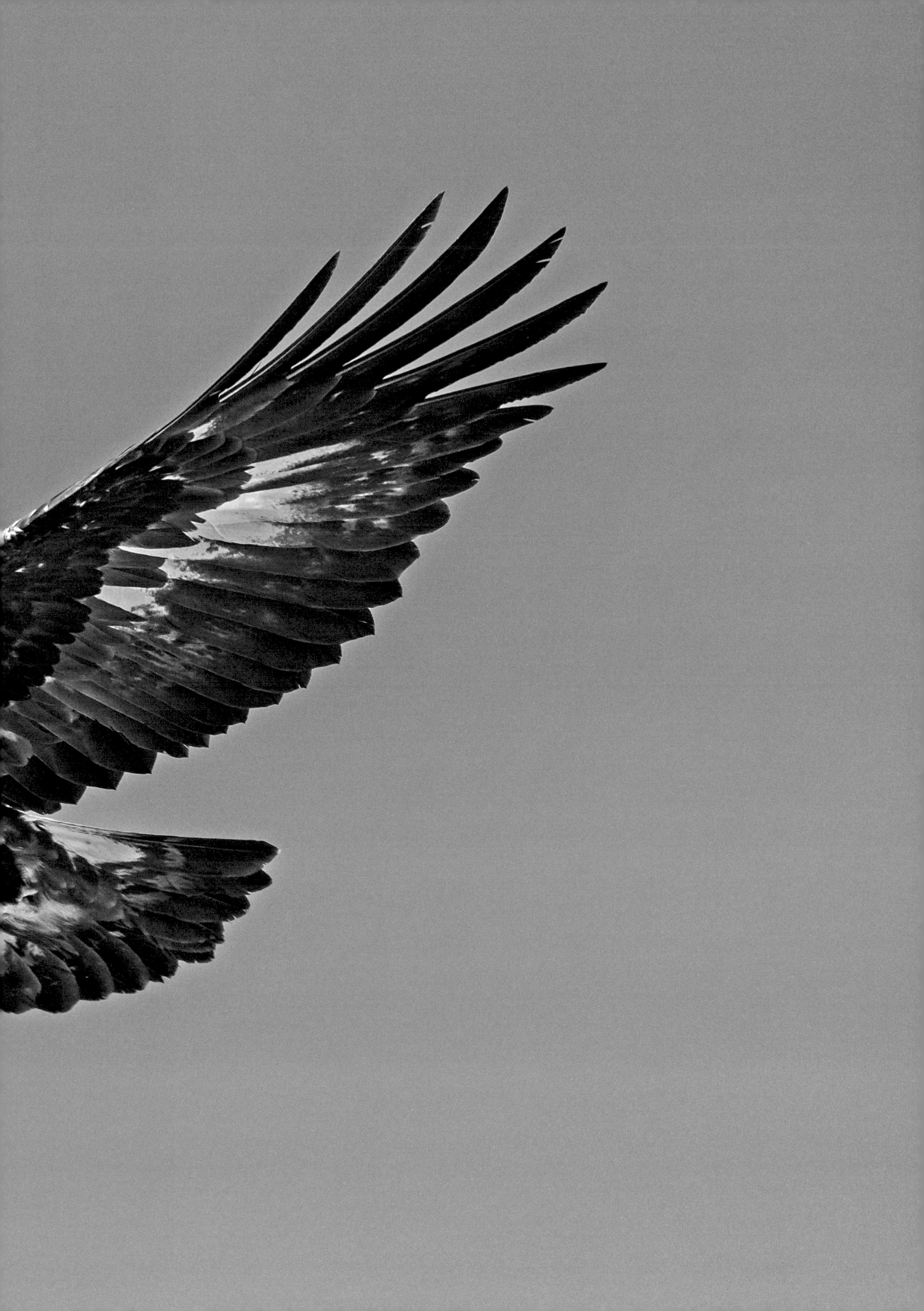

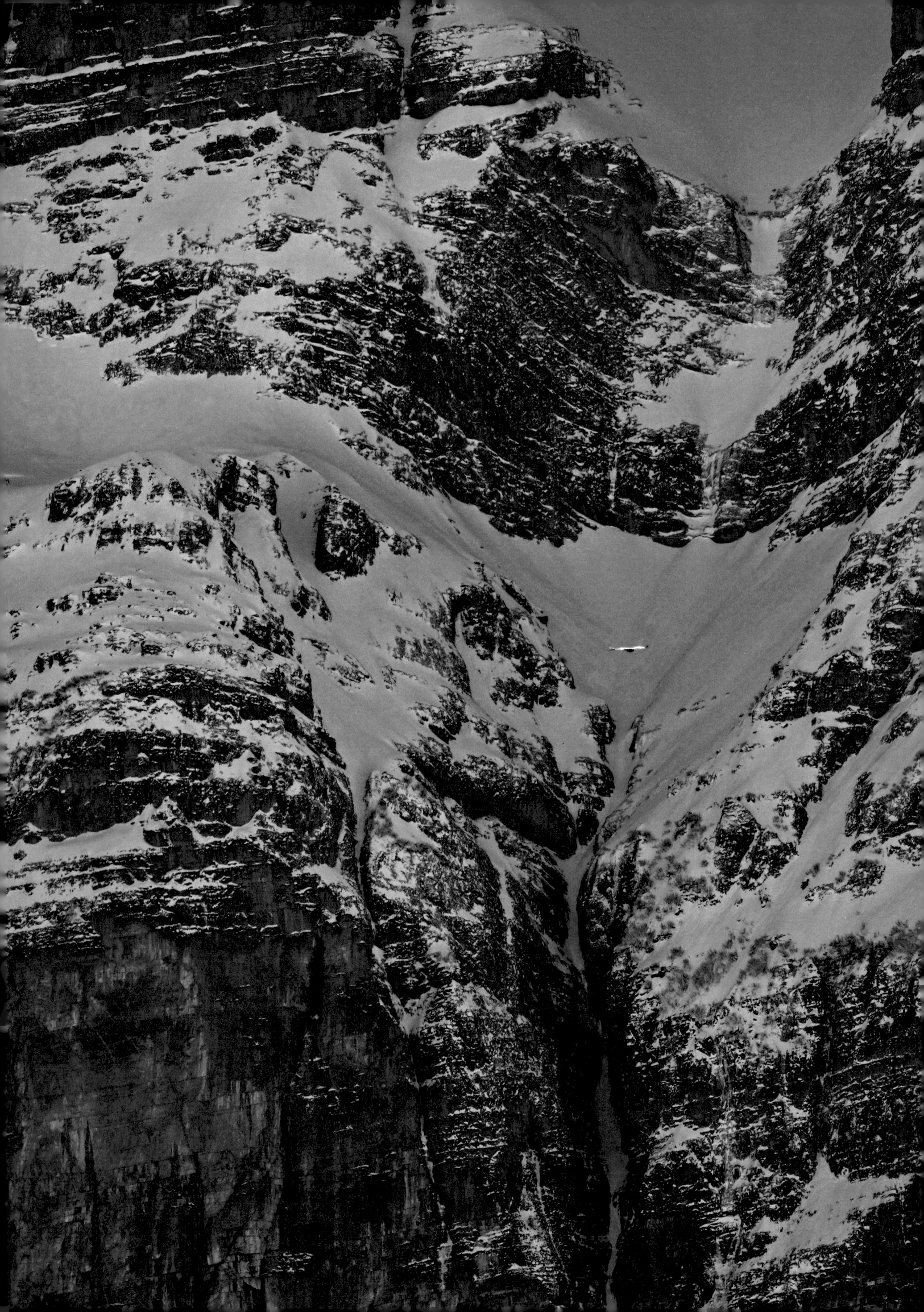

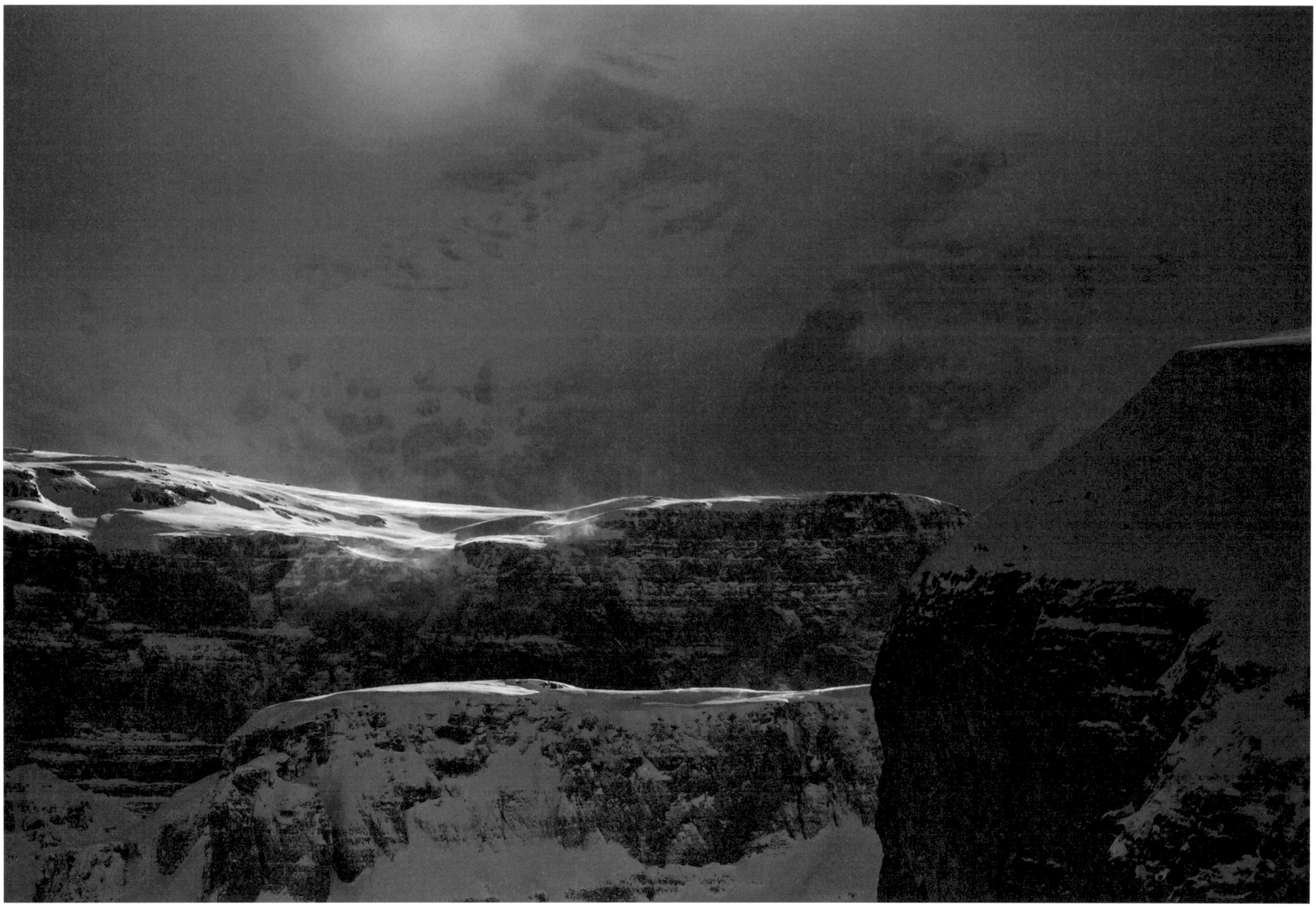

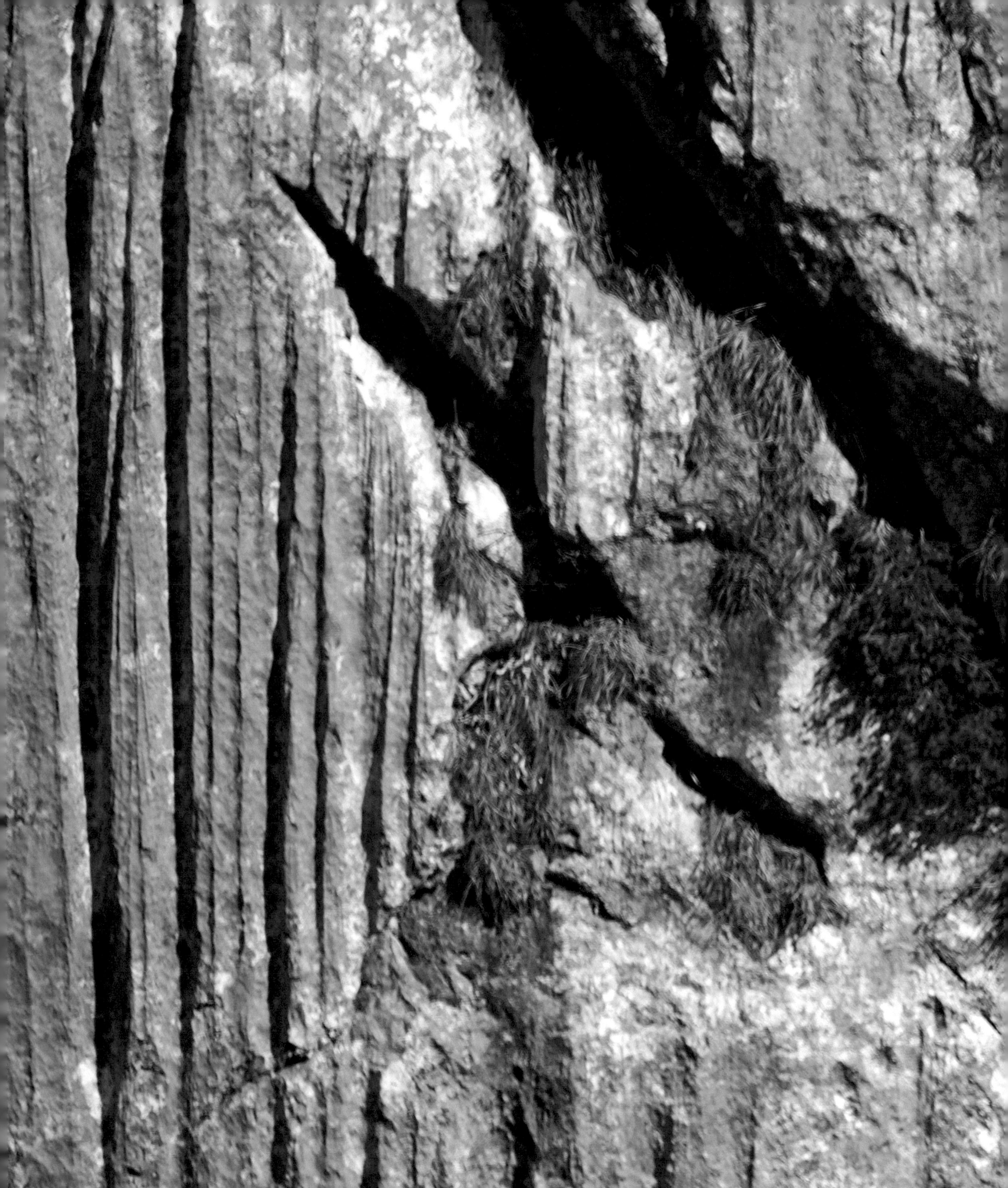

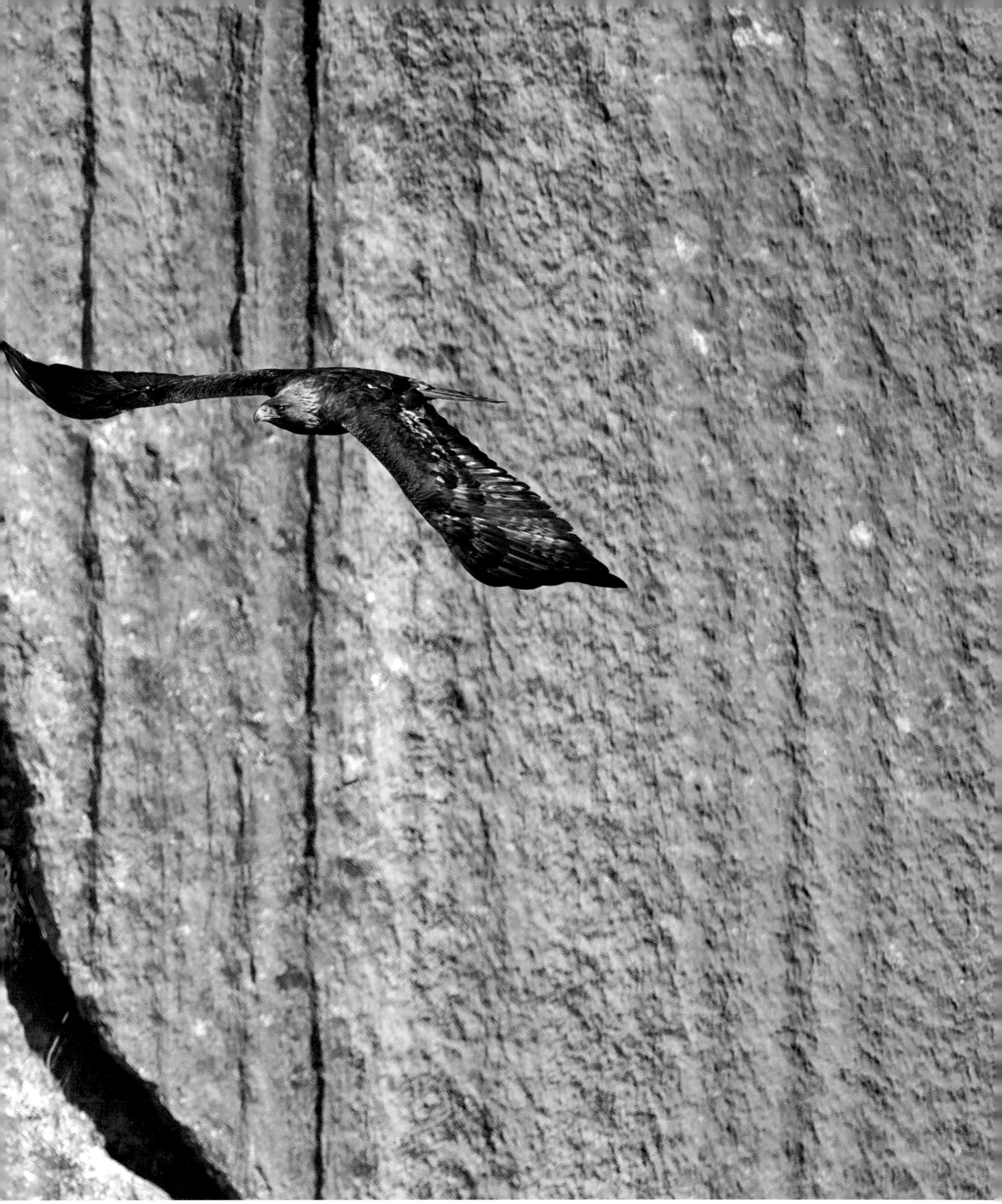

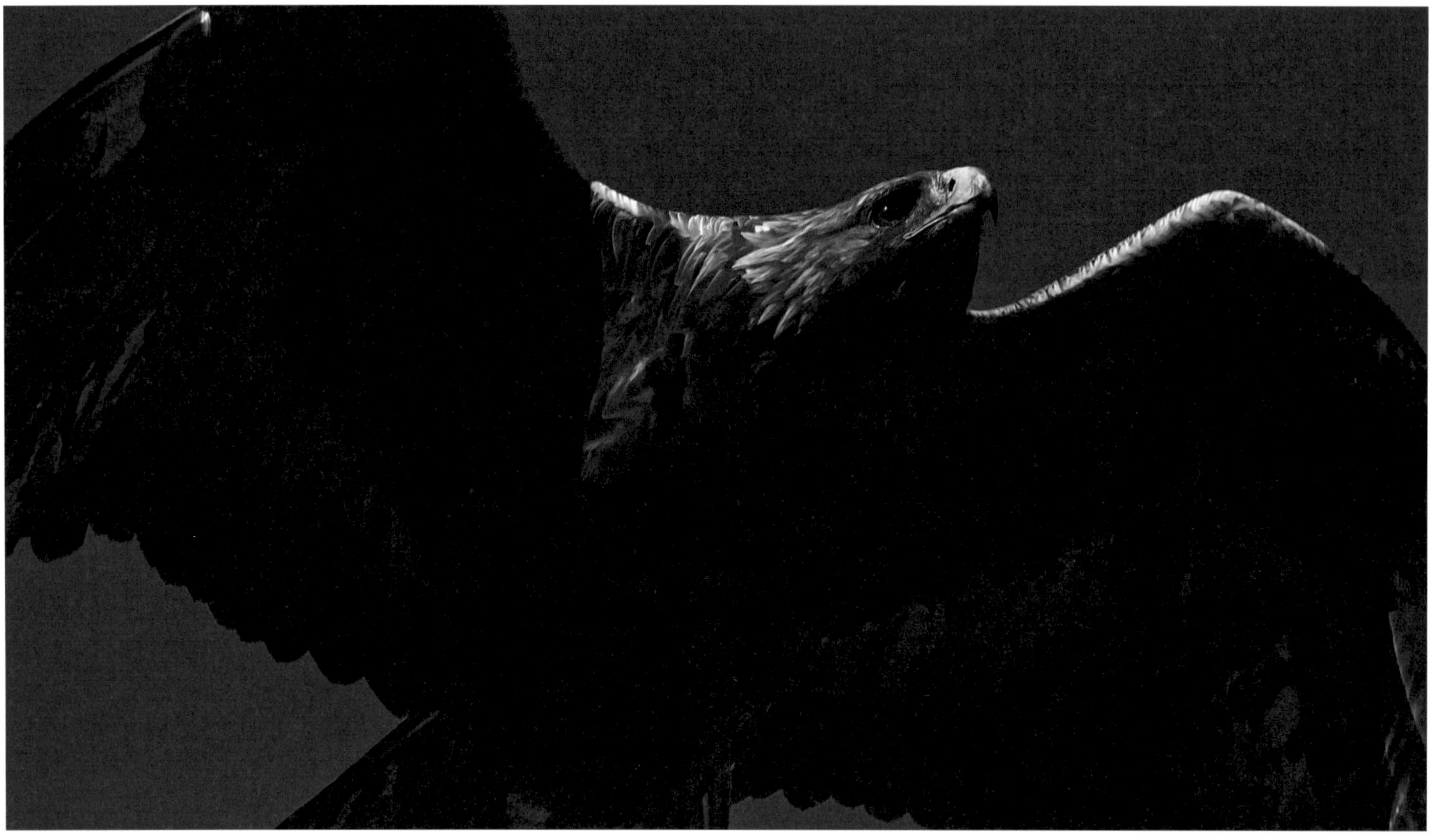

THE EAGLE'S AZURE CROWN

ALFRED LORD TENNYSON \
1809 — 1892 \ 116 \

The Eagle

Fragment

He clasps the crag with hooked hands;
Close to the sun in lonely lands,
Ring'd with the azure world, he stands.

The wrinkled sea beneath him crawls;
He watches from his mountain walls,
And like a thunderbolt he falls.

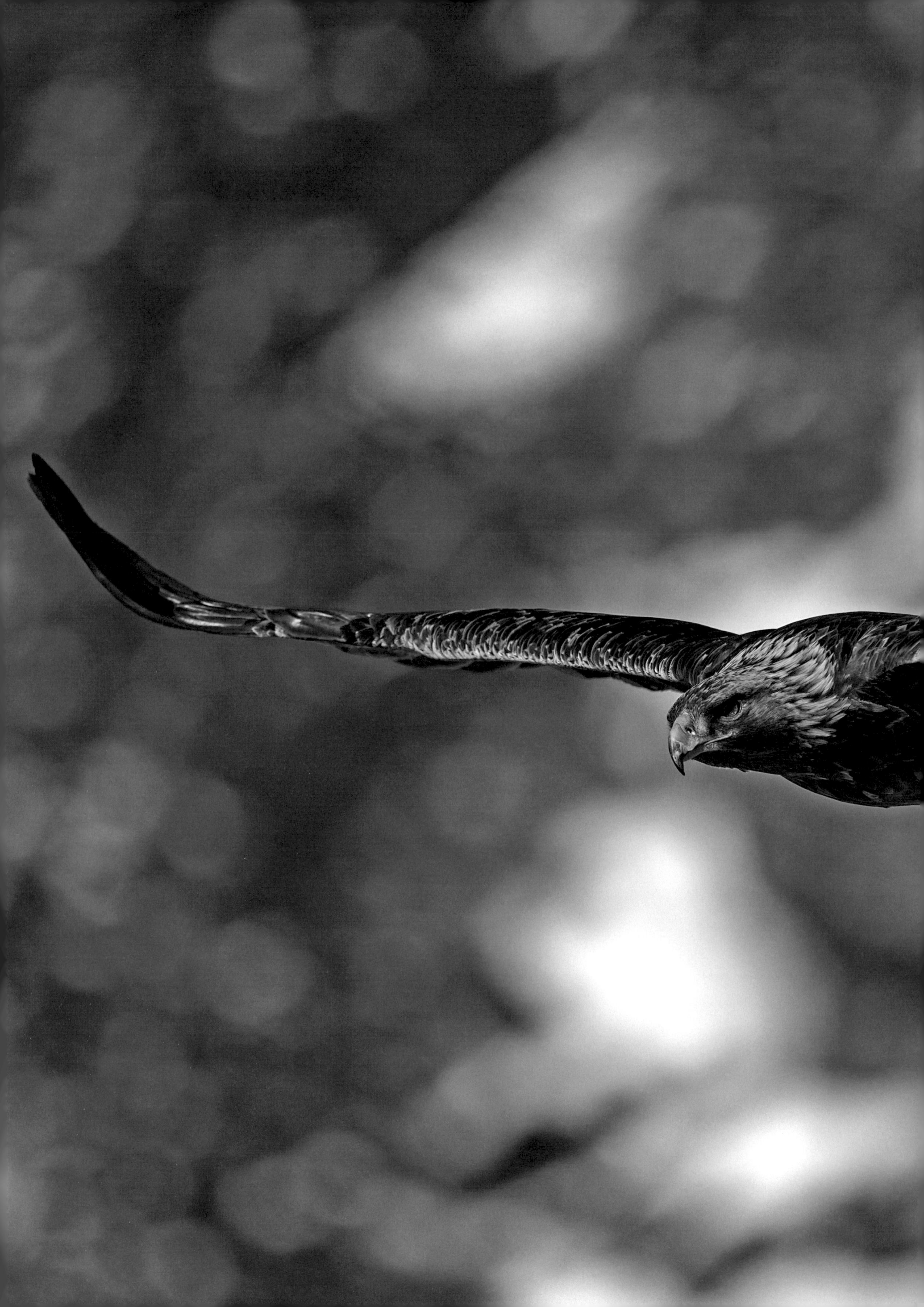

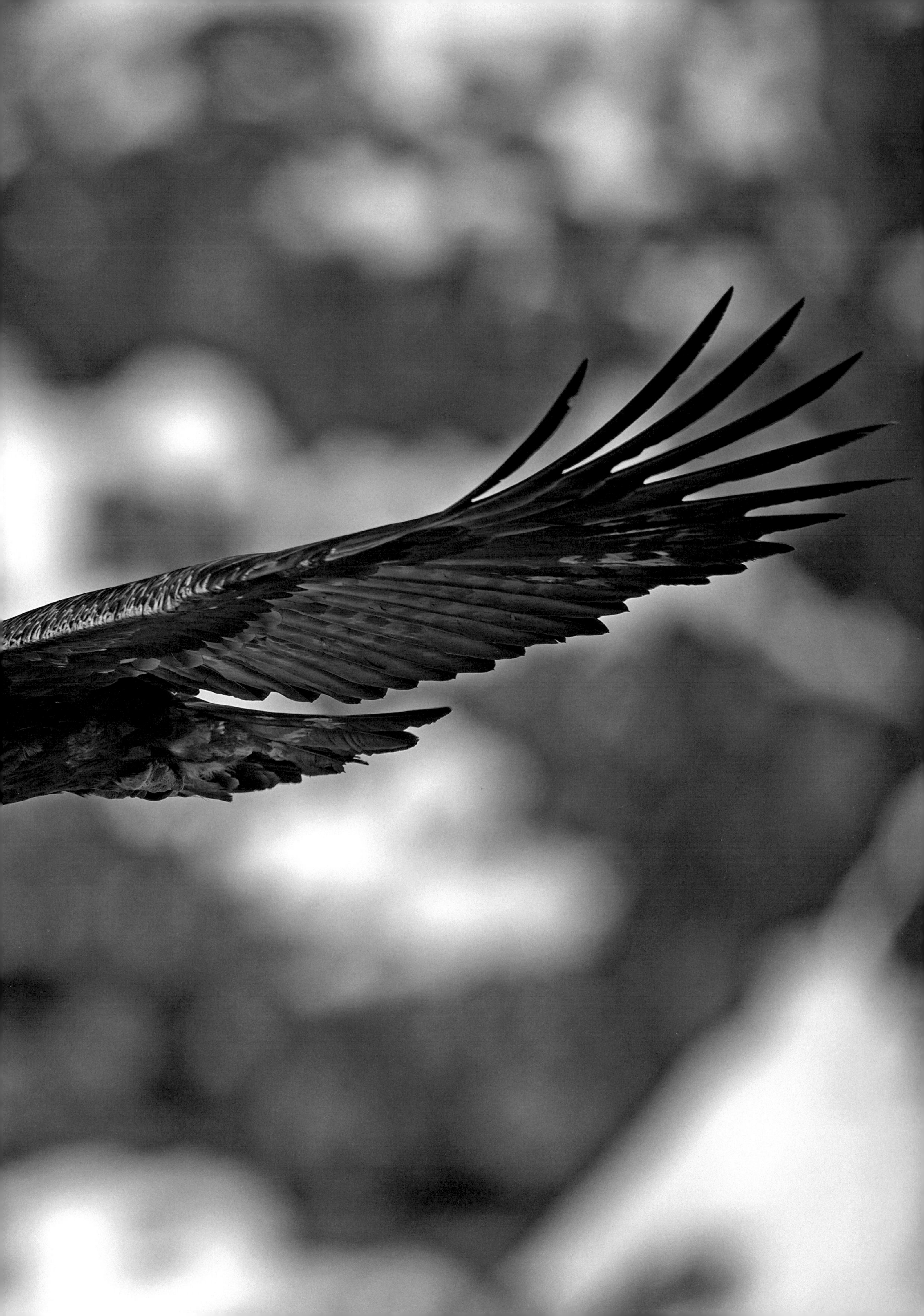

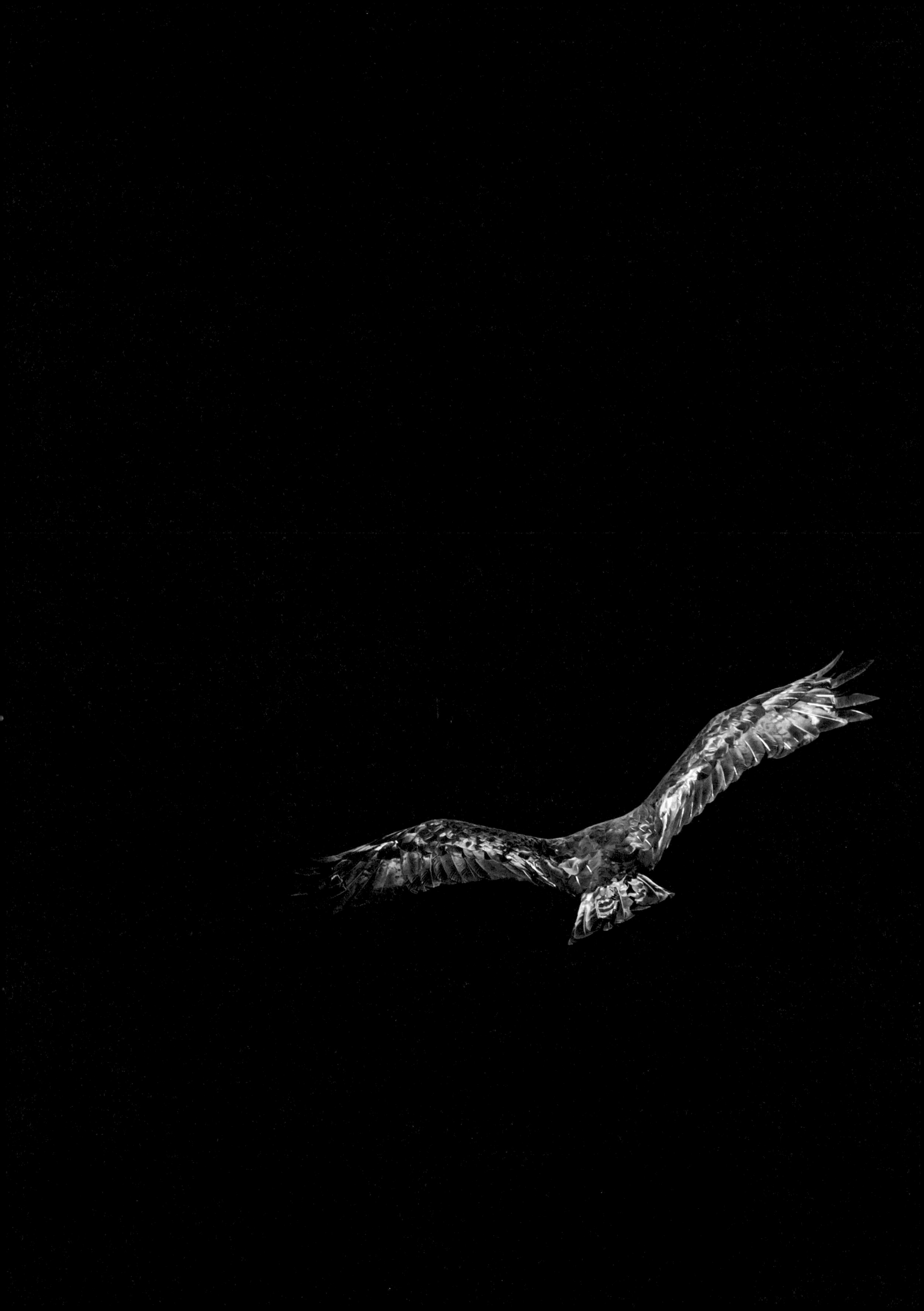

The golden eagle: a challenging subject for photography

If you decide one day to photograph wild golden eagles in your mountain range, I have some advice for you. Plan a good amount of time for your new project! I am not talking about hours, days, or months, but I mean years. Wildlife photography in general is challenging, and the artist needs to be seriously committed to the subject that he wants to study. Bird photography is a special branch of wildlife photography that is well known to demand particularly high dedication from the photography enthusiast. If the bird of interest has its habitat in a difficult-to-access mountain range, uses a territory as large as several dozen square kilometers and is particularly rare, we have the ultimate challenge for the wildlife photographer. Often, captive and trained golden eagles used for falconry are photographed or filmed for wildlife documentaries. If wild animals are documented, they are regularly lured with bait, and the film crew waits in a hide nearby for the breathtaking shot. But this is not what I wanted. I dreamed of obtaining high-quality photographs of native and wild golden eagles in their pristine alpine mountain habitat without using any kind of bait.

When I started the project, I simply wanted a spectacular animal for wildlife photography in my home country, so that the time between photography trips to exotic locations in Africa or India would pass more quickly. I had no idea that this project would keep me occupied for several years and develop from a weekend hobby into a passion, and even an obsession that made me quit my safe job as a scientist in biophysical research.

On innumerable excursions to the mountains of Central Switzerland over the years, I learned where to find the king of all birds and I started to study the golden eagles' behavior and daily rhythm. To photograph these majestic wild creatures, I scouted suitable spots in steep rock walls and explored lookouts where I expected the raptors to fly by. I reviewed the eagles' behavior in the different seasons of the year, studied their change of habits in high or low pressure weather systems, took the local wind systems and daily development of thermals into account, and learned where these birds of prey hunt and how they transport their kill to the eyrie. All this information was needed to maximize the likelihood of getting the object of desire in front of my camera for at least a couple of seconds, to obtain a picture. After years of studying these huge predators, the chance of a golden eagle sighting on a particular day is around 90 percent. Meaning that on nine out of ten excursions, I will spot a golden eagle somewhere, either as a tiny black dot circling one thousand meters above me or, if I am lucky, during a five-meter-close fly by. Obtaining a high-quality photograph on the other hand is a totally different challenge.

At the beginning, I was happy to obtain a picture of a golden eagle at all, and even if the mighty bird was out of focus, just visible as a small dot in the corner of a frame, or seriously under or overexposed, it did not matter too much. I was genuinely proud that I had again managed to find the elusive golden eagle and that I could even document my achievement on a picture. But year after year my standards became higher and I slowly started to think more about the photographic challenge of the project. Suddenly the size of the eagle in the picture started to matter, I wanted the bird crisp and in focus at high resolution, and the dynamic range of the camera sensor for post-processing images became important. It was such an extraordinary challenge to find these birds that I now also wanted to get something back in terms of high-quality photographs. For a day out in the wilderness I needed to get up very early in the morning, drive sometimes two hours or more to the mountain range, hike up to a cliff or lookout with a heavy backpack for another two hours, and then wait for hours for the bird to show up for a couple of seconds or not at all. If I missed the shot due to the limited capabilities of my camera gear, the effort was in no comparison to the meager result. Therefore, I decided one day to get professional camera equipment so that I could only blame myself and not the gear anymore if a picture was not satisfactory. What had started with an affordable, handy, mirrorless APS-C format camera with a full format 70–200 mm f/2.8 lens, turned into a 35 mm “full frame” high resolution sensor DSLR camera with an expensive 500 mm f/4.0 telephoto prime lens. This upgrade not only broke my savings but also meant that my backpack for the steep hikes now suddenly weighed fifteen kilograms instead of the bearable twelve kilograms I was used to. But I was okay with this, thinking that it is worth the effort, and that the tough hikes with the heavy backpack will keep me in shape. Freezing birds in flight demands fast shutter speeds and a fast lens with a wide aperture, and having enough light entering the objective helps enormously. Other technical specifications that really matter to get a high-quality picture of golden eagles are a very fast and accurate autofocusing system, fast frame rates, low noise at slightly elevated ISO levels, and a high dynamic range of the sensor for post-processing the images. A sturdy monopod or tripod to keep the heavy telephoto lens stable is also very helpful, as hand-held shots tend to contain slight motion blur. Only professional camera equipment covers all these points, and if everything comes together the photographs can be of breathtaking quality.

However, moments where the sighting of a golden eagle is ideal to get a great picture are very rare. And if it happens, there are a lot of traps and difficulties that can spoil the chance of a potentially award-winning image. First and most important is the light and the position of the mighty raptor with respect to the sun. Ideally, the sun is at the photographer's back and the bird flies slightly below the observer so that the beautiful colors of the plumage shine and the bird's magnificent golden nape, yellow beak, and hazel eyes are accentuated. If the raptor is above the observer, all the plumage is in the shade, resulting in a black silhouette of the bird where not much detail is visible. The intense light during the summer months, when the sun is at its zenith, creates hard contrasts and deep shadows which often result in difficult to process photographs. This is the reason why I prefer the winter months for golden eagle photography, when the sun is just above the horizon and illuminates the raptor with a soft light from a low angle that creates incredibly vivid and intense colors. This means, of course, getting your snowshoes or skis out and bringing along warm winter clothes to be prepared for a cold day in the mountains. To get the fast-flying bird in focus and quickly take a picture with the right camera settings is often almost impossible. Imagine that, after waiting several hours on a ledge high up in a cliff, the raptor suddenly shoots at high speed around a corner of the rock wall and gives you five seconds to take a picture before it decides to dive deep down into the valley or disappears in another direction. To quickly pick up the camera, find the bird in the long telephoto lens, get the focus point on the rushing eagle, and have all the camera settings right for the position and illumination where the subject unexpectedly appeared is sometimes very challenging. In these moments, a good portion of luck helps to get the great picture, which I always have in mind, but which is so difficult to capture in reality. Another omnipresent difficulty, that can spoil the greatest moment, is thermal distortion, relating to the use of a long telephoto objective. It is ironic that exactly the physical phenomenon of thermals, that enables the king of the skies to effortlessly soar above cliffs and reach the highest altitudes of all birds, also makes photography incredibly difficult. If the sun strikes cliffs or hillsides at a steep angle the air will be heated up quickly and starts to rise fast along the vertical. These updrafts, called thermals, are used by the raptors to soar upwards across rock faces and mountain slopes and enable the bird to reach higher flight levels.

But the fast-rising hot air also creates optical distortions of the field of view, due to temperature differences, which result in a blurring of the photographic subject and yielding a soft picture with reduced definition. Especially when using a long telephoto lens, that cuts through a large volume of air before reaching the subject, this phenomenon can be so severe that pictures often cannot be used at all.

But to get a great picture of a golden eagle, most important of all is patience. One not only needs to wait for hours for the legendary bird to show up, but the wildlife enthusiast should also be in the field as much as possible to increase the likelihood of a great sighting. In case one gets frustrated sitting somewhere on a peak with no eagle in sight, I have a trick that almost always helps bring the raptor close. First one needs to relax the tired arms and put the camera down; then one either digs deep in the backpack for a lunch sandwich and enjoys a meal, or starts to distract oneself with the mobile phone. I guarantee that it will not take more than two minutes for a golden eagle to show up and deliver an incredible display that would for sure result in award-winning shots in every wildlife photography competition. It is just that during the five seconds that it takes to drop the sandwich or put away the cell phone, pick up the camera and aim the lens at the bird to capture an image, the mighty raptor will already have disappeared around a corner of the cliff, or gained so much height that a great shot becomes impossible. The golden eagle is a truly tricky subject to study with a camera, but incredibly rewarding if everything comes together and a great photograph of the huge raptor results!

How it took five years to find the golden eagle's nest

To find the nest of a golden eagle and observe the breeding, hatching, growth, and upbringing of young eagle fledglings by caring golden eagle parents is the ultimate dream of the wildlife enthusiast. In several golden eagle territories, I have always known where to find the eagle's nest but, as those nests lay in designated wildlife reserves, I have always refrained from visiting them in order to respect the law. In other mountainous regions that did not have any official restrictions I could regularly observe and photograph golden eagles, but never ever found any hint of where the eagle's nest could be hidden.

It was on a beautiful day in April in the Canton of Glarus when I finally discovered the eyrie of a golden eagle pair after spending innumerable days over more than five years in the mountains looking for it. On that day, I was strategically positioned on a spacious ledge in a tall cliff where I knew that around midday strong thermals develop which the eagles use to gain elevation for their daily patrols over the valleys and mountains of their territory. As expected, exactly at noon the eagle pair flew along the valley to the bottom of the cliff where they caught the upwind to get lift. The spectacle lasted maybe one minute, after which the predator birds had already gained two hundred meters of elevation, passing my position to fly along the rock wall above me. But this time something was different. I observed the smaller male eagle soar along the cliff maybe fifty meters above me, when I suddenly realized that the eagle had disappeared. Eager to get more photographs, I turned to the large female that was a bit behind and, because of her much larger size and weight, slightly more struggling to get lift compared with her male companion. But soon after I lost sight of both eagles, assuming they both had gained several hundred meters of elevation, and already flown far away. Curious about the sudden disappearance of the first eagle, I walked along the ledge and climbed up the steep slope to the foot of the rock wall above me, in order to check what could have happened. It was exactly at this moment when the male eagle came back, flying from the opposite side along the rock wall. Yet in contrast to anything I had seen so far, the eagle outstretched its long feathered legs, with the bright yellow feet and black talons, as it would get ready to catch prey. The eagle repeated this behavior twice, but then turned away and was out of sight again. Puzzled by this move, I guessed that I had disturbed the mighty bird in his plans, and decided to move away from the place and hide under cover of a small pine tree approximately a hundred meters away. Then, after half an hour, I saw the male approach the same site again, this time flying straight towards the cliff face and coming from the void over the deep valley. Around fifty meters before the rock wall, the eagle angled his wings to accelerate and stretched out his legs, aiming at a small cave in the cliff. But to my surprise the predator bird changed direction just before the cliff, and used the speed gained from the short dive to fly vertically upwards another fifteen meters, and then elegantly touched down on a pile of branches positioned on a small ledge in the inaccessible cliff. "The eagle's nest!" I jubilated. "I finally found the eagle's nest! My god it is so well hidden and hardly visible. This is the place where the male bird had suddenly disappeared earlier in the day!" I thought, and checked the pictures I had taken of the landing eagle. But to my big disappointment, all pictures were out of focus, and I did not have a single good shot from this momentous discovery. As the entire scene from the first sight of the bird to the fast landing took maybe just five seconds, I had misfocused accidentally on the cliff and the mighty predator was just visible as a smeared-out blur. However, the anger about my mistake lasted only a few seconds and I realized what fantastic opportunities the discovery of the eyrie meant for the future of my eagle project. Aware that it would be a highly delicate balance between taking stunning photographs of the eyrie and its inhabitants and carefully ensuring that the eagles would be untroubled by my presence, I was nevertheless looking forward to the coming months.

The irony of the story is that precisely five years before this happened I sat on top of the very same cliff above me and watched a golden eagle fly several times straight towards the rock wall below me and disappear from sight. Curious at the time what the bird might have found there, I climbed down a ravine to reach the foot of the cliff and discovered the cave which I entered and examined. To my frustration, I found neither remains of prey nor a nest in that cave, and I could not resolve the mystery of the eagle's behavior. It took five years to realize that the nest of the eagle pair was not in the cave but on a small ledge just fifteen meters above it.

The curious case of a cord attached to a wild golden eagle

In very early times of the golden eagle project, I had a highly interesting sighting of the king of all birds. At first, I was confused about what I witnessed, and I only understood later, when I looked at the photographs at home, what rare behavioral phenomenon I had been blessed to observe. Sitting on top of a cliff, I spotted a large adult golden eagle female circling approximately hundred fifty meters above the valley bottom in search of an uplift. Knowledgeable of its surroundings, the golden eagle soared over to the flank of the mountain that was struck by warm spring sunrays, in the hope that the developing thermals would provide uplift. The eagle indeed seemed to have found a thermal stream, as it started to circle in tight spirals to get higher. But to my surprise, the raptor severely struggled to get lift, and started to explore several other locations in the valley, with the meager result of only hundred meters elevation gained during the next fifteen minutes. As the scene unfolded, around five hundred meters below me, I took the binoculars to see what was going on. I followed the eagle and thought that the flight pattern looked as if the animal were much heavier than usual. The eagle seriously struggled to get lift and often used a laborious flapping flight mode, which was very unusual. Then suddenly I discovered that the beautiful and large female had a more than two-and-a-half meters long, thick cord attached that it dragged along below and behind its body. “What is going on here?” I thought. Watching the eagle struggle I felt more and more uneasy and suddenly thought that the cord must be the leather strap that falconers attach to the feet of their dressed raptors to tie them to a post. “What idiot brought his golden eagle into this remote valley to let it fly free or hunt?” I thought and got very angry. “This is a pristine home range of a wild golden eagle pair. If they find the intruder a fierce territorial fight will develop, and the animals will get injured or even killed.” I went on in my thoughts. Meanwhile, the golden eagle had managed to soar a bit higher and I took my camera to take photographs. At the time, I only had a small APS-C format camera with a full-frame 70–200 mm F/2.8 lens attached, giving me a maximum reach of 300 mm at F/4.0. “The eagle is still too far away for a good shot with my limited equipment, but at least I will have some pictures of this strange encounter." I thought and took a couple of photos. Then, after another twenty minutes of trying to climb higher, the eagle seemed to have reached its target altitude, disappeared from view around a bend in the valley and was gone.

In the evening, after I had reached home, I quickly downloaded the photographs and looked at the pictures. It took a while to understand what the curious cord attached to the golden eagle was. But when I finally realized what was going on, I almost fell from my chair. The thick cord was not of uniform diameter and had several distinct features. It started with a circa fifteen centimeters long, narrowing piece that was followed by a much thicker and longer bulging part. After that the cord was more uniform over the next two meters, but thorough inspection revealed parts that were twisted or looped out. When I then investigated the part of the cord near the talons, I discovered something very different. Two furry hind legs with black leathery feet each of which, with five toes, were dangling below the eagle's belly. "Oh my god! The eagle had killed an adult marmot that must weigh around three kilograms, and had tried to carry its prey to the eyrie for feeding the young or to another safe place to devour the fat rodent on his own." I realized. "What a beautiful sighting!" I thought. "But then, what is that two-and-a-half meters long cord attached to the dead marmot?" I asked myself, and realized that it was the entire digestive tract of the poor gnawer, including the esophagus, stomach and small and large intestines, all of which were hanging out of its belly. A truly curious case, and definitely an extraordinary sighting.

What if there is no golden eagle in sight: other beastly encounters

Wildlife photography is an excellent patience game that makes anyone who ventures into this demanding subject see the world in a different, calmer, and slower way. There is an omnipresent rush in our fast-paced society for quick satisfaction, for immediate availability of anything we wish, for rapid-fire sensational news with shallow information, or for the prompt fulfillment of all our dreams. In nature, all this is not possible. Nature has its own pace, and the only way to profit from it is to dive in and wait. Although this sounds at first boring and tedious, nature will reward the patient observer with the most incredible wonders, which often are completely unexpected and change the visitor's personality to that of a more calm and grounded human being. Over the years, a state of inner equanimity, tranquility, confidence in oneself, and faith in life can be reached, while at the same time the senses are sharpened to heightened awareness. When you are out in the wilderness and wait for hours for something to happen, vision and hearing will become incredibly sensitive, so that even the most miniscule change in the surroundings will be noticed. After a strenuous hike with a heavy backpack it can easily take several hours of waiting until a golden eagle might fly close enough to the lookout of the photographer to take high-quality pictures. But, on many days, this will not even happen and the object of desire will not show up at all. So what to do with all this time? Well, the mountain habitat of golden eagles is also a playground for many other interesting animal species, and the flora is incredibly diverse and special too. Not to forget the stunning alpine scenery in which everything unfolds, and which provides ample opportunities for fantastic landscape photography. So instead of getting bored and frustrated that the mighty raptor is not showing up, there are plenty of other interesting subjects to experience, study, and photograph. With sharpened senses one needs to investigate footprints and animal tracks, look at small paths or swales in the high grass that could point to nearby ungulates, analyze droppings or feathers on the ground that might hint to a recent meal of a fox or a badger, and listen carefully to the sounds of nature at all levels. It is almost guaranteed that soon after an interesting mammal, bird, or insect will show up and provide an exciting opportunity for photography or observation.

The most common sighting of mammals in the Swiss mountains is definitely the chamois. Especially in the early morning, right after sunrise, and usually during the ascent to a lookout, chamois can be spotted grazing on steep snow-free patches. As they have very sensitive hearing and smell, it is often the gray-white ungulates, with the two pretty dark stripes running over their face, that first notice the encounter and start to withdraw. Sometimes at first one hears only the sound of rolling pebbles on a steep mountain slope or the hissing of a territorial male, before looking into the binoculars confirms the presence of a cheerful female gang or a crabby solitary buck. As chamois change their look along with the seasons, they are always an interesting sighting. It is a great pleasure to watch the rivalry of testosterone-loaded males, well-fed and rounded in November and December, with dark brown and thick winter fur, chasing each other down steep mountain slopes to fight out their mating rights. After a long season of snow and ice, the thick winter coat gets patchy and rough during the spring months, when it is replaced by a short, light brown and gray summer fur with a black line along the spine. With a bit of luck, newborn baby chamois can be sighted in May, or later during the summer months, when the brave miniature versions of this beautiful animal already follow their mothers everywhere they go. As it takes three years for the young animals to reach maturity, there are always individuals of different age and size mixed together in the ten to fifteen-member strong herds of females. Since both males and females possess fine, backwards-curved horns, up to twenty-five centimeters long, the gender can best be distinguished by their social behavior. Females usually form small herds, while males are solitary and only join female groups in the mating season for a short time towards the end of year. Probably the most remarkable feature of chamois is their incredible agility in extremely difficult and steep terrain. It is frightening to observe them stand on tiny footholds in the midst of a vertical drop of several hundred meters whilst looking for the tastiest little herb that might be hidden in a crack of the cliff. If these animals are startled, they can perform astoundingly fast flights down steep cliffs or gravel fields, on which they take long jumps or sprint at high speed quickly out of sight. Similarly impressive is their strength and endurance when they have to escape up steep rocky terrain, or march through shoulder-deep snow in winter. Truly magnificent and perfectly adapted to the harsh alpine habitat, chamois are always a pleasure to spot and observe. In the context of the golden eagle, these mammals are particularly interesting as their young are a major part of the raptor's diet, especially in winter when marmots hibernate and are thus not available.

Although sightings of chamois are by far the most common, with a bit of luck and patience other ungulates can also be discovered.

The proud ibex, with its impressive long horns and its stocky appearance, normally lives above the forest boundary and is a common sighting if the photographer decides to climb higher up into the rough world of rocks and snow, where only small shrubs, sturdy grass species, moss, and lichen survive. Roe deer, on the other hand, live at much lower altitude, so they can best be spotted at the edge of the woods in the early morning when driving to a parking place where the hike will start. The much bigger and powerful red deer is extremely shy and very hard to observe. My encounters with these big animals usually took place in the early evening in winter, during descents back into the valley. At this time of day, herds of red deer descend from the mountains and gather at ice-free rivers in the valleys to drink, before returning to their hiding places in the dense undergrowth of the forest later at night.

Smaller residents of the alpine mountain habitats that can be spotted with a bit of luck and a trained eye include mammals such as the abundant marmot, the rare brown hare or the pretty and completely white (at winter times) mountain hare, the sly fox, the clever badger, the slender weasel, or the bushy-tailed squirrel. A remarkable observation that I keep in mind was an encounter with a fox whom I watched from close for half an hour hunting mice. The healthy and beautiful animal then finally succeeded with an elegant high jump to catch a big example of a fat European snow mole that he proudly carried between his teeth for a couple of hundred meters before he finally devoured it with relish in front of my eyes.

Although the top predators of the Alps, the bear, wolf, and lynx, slowly find their way back into Switzerland, they have not yet arrived in large numbers in the central part where I usually look for golden eagles. The closest I came to one of these mighty hunters were footprints of a lynx that had obviously carried out an attack on a grouse, as an entire snow field was covered with pugmarks of the fierce fight between the two opponents. Who won remains a mystery, as I did not find evidence of the lynx plucking the bird's feathers before its meal. If no mammals can be seen, there are always birds around that often present wonderful opportunities for observation or photography. Avian species that share the habitat with the mighty golden eagle can normally be seen on every trip. It is always a great pleasure to observe the highly territorial raven pair that is common to the valleys and mountains of Switzerland. Ravens are among the most fantastic flyers in the avian world. They are usually the first birds in the morning hours that use developing thermal updrafts to gain height along cliffs. Effortlessly, they soar without flapping their wings and easily climb up in circles when thermals are still too weak for the heavier eagles. As they form pairs for life, male and female always show up together and keep in touch with each other with short calls. Often, the far-carrying call can be heard first in the valley, even minutes before the birds appear. Another peculiar behavior is that they regularly roll onto their backs with fully folded wings for a short time and call before they turn back into normal flight position. Why these intelligent birds do this flight maneuver, repeatedly and for a long time, is an unsolved mystery. Ravens are highly territorial and do not tolerate a golden eagle entering their air space, although they share the same territory. The mighty raptor, although much bigger, will immediately be attacked and followed for many minutes all along the valley if spotted nearby. Such attacks can result in spectacular photographs when the large predator bird finally has had enough of the annoying black raiders and flips onto its back to try to snatch the assailant with its fearsome talons. Due to their aggressive territorial behavior, I always refer to the beautiful birds with their shining black feathers as the police of the mountains. If ravens are around there is a high chance of a golden eagle sighting, as they will spot the king of the sky first and immediately start a noisy raid to chase away the much larger and heavier eagle.

Large flocks of yellow-billed chough seem to have an inner clock, since they often appear exactly at the same time in the afternoon, exploiting upwinds along the cliffs to gain elevation, before soaring to an opposite side of the valley. In constant chatter with each other, these small birds are excellent flyers and a great pleasure to watch. If they appear in smaller groups or in pairs, they can be very curious and will stay close to humans for a while to inspect what is going on, or in the hope of a bite. They often fly within centimeters of rock walls and execute acrobatic turns to chase each other down a cliff or valley at very high speed. Very social and playful, these birds always induce a feeling of envy in me as they seem to live such fulfilled and happy lives, always surrounded by family and friends.

Another legendary bird that is often found sharing the habitat with the golden eagle is the peregrine falcon. With a profile resembling a fighter plane, a stocky, barrel-shaped body and short pointed wings, the fastest bird on the planet soars without difficulty along cliffs, always scanning for another bird it could attack and kill. Matt gray from above and beautifully black-white striped from below, the noblest of all falcons can easily be identified by its feather-covered legs with yellow feet, yellow cere and eye-rings, and deep-black eyes with a very distinct black beard under the eyes. It often follows the golden eagle at slightly higher altitude, probably in hope of getting its share, if the mighty king catches prey. With some luck, the observer might witness the famous nosedive of the large, crow-sized falcon during an attack on another bird. As soon as the raptor has prey in its crosshairs, it will fold its wings to its body, and stoop down like a missile from far above the innocent victim. During these attacks, the peregrine reaches speeds of more than three hundred and fifty kilometers per hour, during which it still processes every movement of the target, and accordingly corrects flight position and trajectory. Upon the violent impact, the victim will be taken totally by surprise and likely killed instantly. With an estimated population of only around three hundred breeding pairs, the peregrine falcon is a rare and endangered species in Switzerland. This makes it even more rewarding when it suddenly appears together with the mighty golden eagle, and both mystic raptors can be observed simultaneously.

Another member of the *Falconidae* is the common kestrel, which can easily be sighted between March and September, when this pretty, small falcon is back in Switzerland before it leaves again for its winter quarters in Southern Europe. With a distinct shrill voice, kestrel pairs constantly communicate with each other or with their young during the breeding season and can therefore be readily discovered. Although tiny compared to a golden eagle, they will also immediately attack the mighty raptor if it ventures too close to the falcon's nest or when their naïve young birds learn to fly under the supervision of their parents. With noisy chatter, the rufous-backed falcon will courageously pester the huge eagle, as long as it takes to annoy the intruder sufficiently that it leaves for more tranquil airspace.

A rarer sighting is the northern goshawk, a powerful raptor that can be seen from above, flying over treetops and using rock faces to gain altitude. Of silver-gray color on top and with a white-gray striated belly and underwings, the agile hunter is easily identified by its unique, piercing, orange-colored eyes and the yellow legs and feet. Its wings are broader and more rounded at the tip, when compared with the peregrine falcon, its body size and wingspan are bigger and, most importantly, the beard typical of falcons is missing. The much smaller Eurasian sparrow hawk has featherless legs, yellow eyes and an often rufous-colored patch on the head and belly between the grayish top and the striped underparts. It likes to join flocks of other birds that circle in thermals, but will then suddenly attack the unsuspecting companions, grab its meal, and disappear.

The Eurasian buzzard can be seen occupying the lowest lying territories in the valley, when one looks down from a lookout high up on a rock wall, waiting for an eagle to fly by. By far the most common raptor in Switzerland, it can be sighted almost everywhere, and is often noticed after hearing its distinct call. It is a bird of a good size, of light brown color that often has a contrasting white band on the underside of its primaries and secondaries. While looking for mice, it soars for hours above fields and often gets into fights with a gang of crows that always try to chase it away. Especially in spring, these birds are fascinating to watch when, during mating season, the male tries to impress its partner with stunning acrobatic displays.

The elegant red kite can also be seen soaring over alpine meadows at the foothills of the Alps during spring and summer, when it is back from its winter quarters in Southern France and Spain. The large but slender bird of prey can be identified by its long, bent wings with distinct white patches on the underside, by its rufous-brown colored body and top side of its wings, the long, slightly forked tail, and the silver-gray head. A rarer find in the mountains of Switzerland is its relative, the black kite, which sometimes ventures into the foothills of the Alps, but is usually more bound to the lakes and rivers of the Swiss Midlands or the hills of the Jura in Northwestern Switzerland.

The bearded vulture is the only bird of the Alps that tops the size and weight of the golden eagle. It reaches a wingspan of more than two-and-a-half meters and can weigh up to seven kilograms. Although the bearded vulture became extinct in Switzerland towards the end of the 19th century, its population numbers are slowly growing again, due to an international reintroduction program that was started in 1986. With a population of only around fifteen pairs, this mighty raptor is very difficult to observe, and despite having spent hundreds of days in the mountains of Central Switzerland sighting wildlife, I was never lucky enough to meet the famous bone cruncher. Nesting sites are known in the mountains of Leukerbad in the Canton of Valais or in the Swiss National Park in the Grisons and these are, therefore, better places to observe this magnificent raptor that sometimes can be seen in territorial fights with the mighty golden eagle.

Depending on the season, there are dozens of other bird species in the Swiss mountains, providing ample opportunities for photography and observation. Among the eight woodpecker species found in Switzerland, the black woodpecker, the great spotted woodpecker, the European green woodpecker, the Eurasian three-toed woodpecker, and the white-backed woodpecker, all share parts of their mountain habitat with the golden eagle. These colorful birds are best observed during a hike through the forest, or open woodland, on the way up to the rocky lookout. While the woodpecker's courtship drumming can be heard for miles in the mating season, it is usually the large dispersal of bark and wood debris around an old tree that reveals their presence. Waiting in a hide nearby will almost certainly guarantee an encounter with one of these busy lumberjacks. Deep holes in the large dome-shaped mounds of red wood ants, on the other hand, point to the work of the European green woodpecker, whose diet consists mainly of these social and busy insects.

While ravens play a policing role in open air space near cliffs and above treetops, it is the Eurasian jay that warns its cohabitants of anything that enters the forest. Rather annoyingly for the wildlife enthusiast, this colorful bird will announce one's presence immediately and instantly scare away potential photographic highlights. However, some inhabitants of the dense forest are so secretive that not even the warning call of a jay makes them leave their hideouts. The reclusive hazel grouse, for example, waits in its hideout to the very last second before taking off from the undergrowth with a loud flapping noise, just a few meters away from the hiker, to reach safer grounds.

Finally, having arrived on top of a lookout, waiting for the king of the sky to show up, I greatly enjoy observing other bird species that love the rock faces as much as I do. It is always a great pleasure to discover the highly specialized wallcreeper bird, climbing up cliffs to skillfully search for insects in small cracks and crevasses of the rock, with its long and fine bill. At first, this gray little bird might look rather dull to the inexperienced observer, but when it takes flight a totally unexpected explosion of colors, rare for a European bird, will delight the nature lover. Crimson-colored wings on a black background, decorated with white dots, make this bird one of the prettiest inhabitants of the Swiss Alps. Enthusiasts for high speed and agility will definitely find pleasure watching the common swift and Eurasian crag martins dart along cliffs in the chase for insects at mind-boggling speeds, whilst performing incredible turns that one would have thought incompatible with the physics of flight. Their fast and erratic flight patterns greatly challenge the skills of photographers, and their hardware, when targeting a high-quality picture. Nonetheless, they are a great joy to watch, and I am always happy when they return from their winter quarters in Africa, ready to entertain.

During the descent into the valley, after a long day high up in the mountains, one is always accompanied by a great variety of additional bird species. They might prefer lower altitudes, inhabit different vegetation zones, have a different daily rhythm and only become active in the late afternoon or evening, or show a little more courage, presenting themselves to the tired explorer. From crested, coal, or blue tits to ouzel, fieldfare, nuthatch, tree creeper, magpie, wagtail, alpine accentor, wood pigeon, and cuckoo, or a rare migratory bird on its way north like the black stork, there is ample opportunity to discover the great variety of avian life of the Swiss Alps. Contented from a day out in the wilderness, it is then often the mysterious call of an elusive owl that accompanies the tired hiker on his last steps back to the car, bus stop, or train station down the valley.

The luckiest person in the world

I was already deep into my golden eagle project, when the institution I worked for organized a scientific retreat in the Engadin. This beautiful valley lies deep in the mountains of the Canton of Grisons in the eastern part of Switzerland, combining the snow-covered peaks of the more than four thousand meters high Bernina massif with several deep-blue lakes that together merge to form exceptionally beautiful scenery.

As part of a team-building event, we went on a hike during a free afternoon and took the funicular from Punt Muragl near Pontresina to Muotas Muragl at 2454 m above sea level, from where a pleasurable walk to Alp Languard started. But not even halfway through, I thought that this hike was too easy for my taste and I decided to leave the group and walk up to the Chamanna Segantini, located near the summit of the Munt de la Bes-cha, the small hut where the painter Giovanni Segantini died in 1899. But when I just branched off from the main path, a colleague that I did not know at the time asked whether she could accompany me up to the hut. I agreed and we hiked up the slope through the fifteen centimeters of fresh snow from the previous night, while having lively discussions about work and our passion for the mountains. When we reached the top, we were pleased that the small hut was still open and served drinks and snacks, although it was already late September and the main tourist season was long over. We enjoyed our little meal and had a look at a book with old photos from the life of the famous painter who had sadly passed away at the young age of forty-one years, at this beautiful place overlooking the lakes of the Engadin and the Bernina massif, while trying to finish a painting. But suddenly, I discovered a moving object in the air below us and stood up to see what it could be. “Oh, look! A golden eagle: what a rare sight!” I explained to my colleague, and added “How lucky you are to see the king of the sky. This is an exceptional creature and very difficult to spot.” “Yes indeed. This is a really big bird and a nice sighting.” My colleague replied, without seeming overly excited. “Lucky her. But she does not seem to know what a privilege it is to see a golden eagle from up-close like this.” I thought. Even the couple that managed the little restaurant came out of the small hut after they had heard my joyful comment and were very pleased to discover the mighty bird visiting their place. Delighted, we all watched the eagle circling to gain elevation along the slope of the mountain before soaring in straight flight to the opposite side of the valley and disappearing from view. As I had just met my work colleague on that day, I did not say anything about my passion for this raptor, and that I spend a good amount of my free time following it with the camera. Later in the evening, after we had descended from the mountain and had reached Pontresina, we agreed that we had spent a wonderful time away from the crowd and promised to contact each other for another hike later in the year.

Several weeks later we got in touch and decided to hike in the mountains of the Canton of Schwyz in the central part of Switzerland. We followed a steep route off the beaten track of mount Fluebrig and reached the summit at around two in the afternoon. We congratulated each other for the successful climb and were eating our lunch, when another hiker who was present on the summit suddenly shouted. “Look over there. A golden eagle!” And indeed, there was a large golden eagle which majestically glided maybe two hundred meters below us, towards another peak in the massif. Quickly, he took several pictures with his small digital camera and proudly presented them to us after the bird had disappeared around a rocky cliff. Although he had only captured the eagle as a small speck, due to the limited range of the lens, he zoomed into the frame and the eagle suddenly filled out the entire small screen. “Look how beautiful he is.” He proudly said and murmured to himself. “I got a picture of the mighty golden eagle.” “Hey, you are really fortunate to see a golden eagle yet again today. You know they are very hard to sight.” I repeated my words from our first hike in the Engadin and thought that she might make a good companion for my photography adventures, having been so lucky to attract the king of the sky.

Just before Christmas, although there was already a good amount of snow in the mountains, we decided to do our last hike of the year. We planned to do a round tour in the Pragel pass region of Central Switzerland where I used to spend a lot of time climbing in my earlier life. It was a very cold day, well below zero degrees Celsius, and the gloomy overcast sky above us was not very inviting. Nevertheless, we left the cozy warmth of the car and started our hike in the early morning. After one and a half hours, tired from the ascent in the deep snow, we decided to take a rest at the foot of a large rock wall. We installed ourselves on a little wooden bench and started to eat our snacks, when suddenly the clouds parted, creating a gap for the rays of the low winter sun to warm our bodies and souls. The south face quickly warmed up a little, and we enjoyed the break and our small bites while chatting about work and life. But when I looked into the sky, I could not believe what I saw. “My god. How lucky can she be? There is a golden eagle approaching the cliff above us, through the opening in the clouds.” I thought, and told my colleague to watch it. “Aha, another golden eagle. They seem to be very common here.” She commented with a wink, while admiring the bird soaring along the cliff. “Not really. They are actually very rare. There are only something like 350 golden eagle pairs in the entire country.” I explained, and suggested we pack up and continue our tour, as the clouds became thicker and darker with every minute. Indeed, the weather got worse and we completed our round in thick fog, reaching the starting point in the late afternoon, tired and chilled to the bone. It was only then, over a hot chocolate in a small restaurant close by, that I revealed my personal obsession with golden eagles and mentioned my photography project. I tried to explain how difficult it was to spot the mighty king of the skies, and that it needs a good amount of patience and luck. “You have no idea. But you definitely are the luckiest person in the world to see a golden eagle on every hike.” Later, I understood that the appearance of the mighty golden eagle on every hike with this wonderful colleague was a happy omen, as she became a very good friend of mine.

Photographing golden eagles high up in cliffs can be dangerous! As I love the intense colors of the winter light, I could recount many critical situations in deep snow with avalanches thundering down left and right. But here comes a different story. Since an eagle flyby might last just a few seconds after hours of waiting, things can get pretty hectic when the object of passion is in sight. Once, not even a minute after I had reached my lookout high up in a cliff, a superb golden eagle appeared, gliding along the cliff below me. I was standing right on the edge of the vertical, three hundred meters drop when I quickly pulled out my camera from the backpack to get ready for the shot. But a second later, and with just the expensive camera in my hand, I watched the backpack free-falling the three hundred meters tall cliff, terminating the fall with a violent impact on the ground, and rolling another several hundred meters down a steep slope before finally coming to a stop. What followed was not only a missed shot but also a tiring, four-hour trek around the mountain to collect the backpack, whilst carrying the heavy camera.

Throughout history, golden eagles have had a fearsome reputation, and were often blamed for seizing lambs, goats, dogs, cats, and even small children, and transporting the prey back to their nests to feed on. In cultures worldwide, this fear has been expressed in fairy tales, found its way into mythological traditions, or played a role in oral and written records where the large bird of prey swoops down onto the land of farmers to carry away livestock or infants. I always wondered whether these narratives were just myths that put the wild hunter in a bad light, or whether golden eagles really were capable of snatching and carrying away a small child of five to ten kilograms. But when I wrote the chapter about hunting strategies and feeding, I stumbled over spectacular footage of golden eagles in Spain that regularly attack and kill young or, astonishingly, even mature chamois, transporting the heavy prey to the eyrie. I imagined witnessing a similar scene one day in the mountains of Switzerland and wanted to document the incredible performance of an eagle carrying a chamois in his talons in a photograph. I knew that golden eagles regularly hunt young ungulates which, with a weight of around five to twelve kilograms, can be transported downwards to the nest. I therefore decided to spend time in May and June in the mountains, when many newborn chamois accompany their moms in the rugged terrain, making them an ideal target for the golden eagle. Although the harsh light from the sun, almost at the zenith, is not ideal for photography at this time of year, I hoped to observe and document the iconic moment of a golden eagle carrying ungulate prey. Already on the ascents to my lookout I regularly observed small herds of chamois in which almost every female had a little baby at its side. "It's the right time." I thought and installed myself in a hide overlooking the valley. The place was close to a golden eagle eyrie, but definitely far enough from the nest not to disturb the birds. "If the resident golden eagles transport prey to their nest, they have to pass my position and I will be able to take a picture." I hoped. Soon I realized that my strategy seemed to work, as there was a lot of golden eagle air traffic towards and away from the nest. Usually during the morning hours, the eagle pair was busy with nest maintenance and patching. For this construction, the birds carried a wide variety of cargo in their claws, ranging from fresh pine tree branches with needles and large bare sticks, to immense tufts of golden-colored dry grass for padding their home. Often, I spotted undefinable, rather large balls in their talons that, after image processing and careful inspection of the photographs, turned out to be pieces of carcasses. I often discovered fury bits, blood, parts of the intestine or other organic material, probably tendons, protruding from the meat ball. Sometimes the eagles transported these lunch balls, as I called them, into the nest and then, just a few seconds later, depart again with the meat ball in their claws, likely to bring it to another eyrie or maybe consume it on a perch. Several times, I was very fortunate to photograph golden eagles with a fresh kill in their talons. A significant change in the flight style of an approaching eagle, a bulky mass under the belly, or legs and feet that dangle down into the air, are definitive signs of a golden eagle carrying prey. Since encounters with eagles happen in the blink of an eye and at high speed, it was often only later, after inspection of the photographs at home, when I recognized what the eagle had killed. Marmot and hare, for example, are the most common prey and about the same size, but distinguishable by their fur color and the shape and texture of their legs and feet. Marmots have a leathery black sole on a five-digit foot with clearly discernible claws, while hares have more fury soles and four large toes on their hind legs.

One day, using my binoculars, I sighted a golden eagle circling in a thermal updraft that had developed above a mountain slope on the opposite side of the valley. The bird was around two kilometers away and slowly gained elevation. From previous observations, I knew that the resident golden eagle pair often used the very same vertical air column to reach higher levels above ground, in order to approach the eyrie on my side of the valley in a straight, slightly descending flight. Indeed, this was exactly what happened. Once the raptor was significantly elevated compared with my position, it left the thermal tube and flew straight over the valley floor towards me. But this time something was different. The bird held its wings slightly bent, so that it glided at rather high speed. It was obviously struggling to keep height, as it rapidly lost altitude. When it came closer, I realized that it had something heavy in its talons that made the fast, straightforward flight laborious. Nevertheless, the golden eagle seemed very determined and continued its approach to the difficult-to-land nest, located in a small cave amid the large rock wall. Then, maybe a hundred meters before the cliff, where the difficult landing procedure would need to be initiated, the mighty bird realized that it was flying too low to reach the nest. The bird sharply turned to the right and continued flying along the rock wall.

It was at this moment when I counted four legs that dangled down into the air under the raptor's body. "An ungulate! The eagle just tried to carry an ungulate to the eyrie!" I was totally excited, but a little disappointed that the eagle had missed the landing in the nest with the heavy prey, at the very moment that I wanted to capture it with the camera. I then followed the path of the golden eagle that, after a few rounds back and forth along the rock wall, flew again straight over the valley bottom to the opposite side where it reached, in a triangular flight path, the very same thermal it had used before. "Oh, it will try again!" I thought, delighted, and hoped to get my shot this time. The eagle obviously knew that it needed to get a little higher in the thermal column at the second attempt. It circled longer in the updraft and started the straight flight over the valley from much higher altitude. In an almost identical flight, the eagle glided at elevated speed with angled wings over the valley towards the nest. Again, I aimed my camera at the eyrie and looked through the eye piece to capture the landing with the kill. But the mighty bird never appeared in the viewfinder. I looked up from the camera and realized that the eagle had missed the landing yet again and had initiated another touch-and-go maneuver right before the cliff. I guessed that the landing with the heavy load in its feet must be very tricky and managed this time, at least, to get a couple of pictures of the bird flying away from me. I followed the eagle with my binoculars, and was relieved to see that it executed the very same triangular flight path, using three different updraft systems, in the hope of completing the landing and successful transporting the heavy load to the nest in the next attempt. "It gives me another chance. How lucky I am today." I thought and got ready for the shot. This time I tried to glance at the bird from the corner of my eye to follow its flight path, whilst simultaneously aiming and focusing at the nest to get the picture I dreamt of. But soon, I realized that the golden eagle also struggled to deliver its kill to the nest at the third attempt. It tried to approach its cozy home and all looked good this time but, shortly before landing, gusts from the strong northerly wind made the flight very unstable and the raptor had to turn away again to avoid a crash landing on the highly difficult-to-land airport. Again, I took a few pictures of the bird with its ungulate prey and discovered that it carried a chamois baby in his talons. Two weeks later, I discovered a snow-white baby eagle in the very same nest and guessed that the golden eagle must have abandoned its tricky landings with the heavy prey to avoid damaging the egg with his offspring. What a considerate parent!

After I had finally discovered an eyrie of a golden eagle pair in the Swiss mountains, I dreamt of documenting the little hatchling's transformation into a courageous juvenile golden eagle that starts to explore the immediate vicinity on its first flights. With dead branches, twigs, and camouflage nets I constructed a tight hide in a group of small conifers, so that only the front glass of my lens remained visible from outside. From this position, I had a clear view up to the eagle's nest that lay in a small cave amid an overhanging cliff at around hundred fifty meters distance. Using a 1.4 times tele-converter on a 500 mm lens, in combination with the high-resolution sensor of my camera, allowed me to obtain reasonably sized images from the relatively distant hide. As I knew that golden eagles can react negatively to disturbance of the nesting site and the surroundings, I really wanted to make sure that I was well hidden and far enough away not to disturb the golden eagle pair in any way in their breeding activities. In addition, I decided to visit the nesting site only fortnightly to minimize my presence. With great enthusiasm, I started my documentation of a golden eagle nursery in mid-April. Already on my first two visits, I felt blessed to observe a lot of golden eagle action. Both parents visited the nest several times, and deposited twigs, foliage, or bloodied parts of prey on each landing, enabling great behavioral sightings. But despite my joy about the interesting observations, I started to doubt whether I had chosen the right site. “Shouldn't the female be brooding her eggs most of the time in April? Why do I observe both male and female on territorial flights, hunting, or collecting material for nest maintenance? What if this is just one of the several eyries in their territory that is regularly maintained and used, but this year's nesting site lies in another place?” I asked myself but continued the observations through May in the hope of seeing hints that the nest was actually that year's fledging place. But still, both eagles spent a lot of time in the air, and I therefore concluded that the resident eagle couple were either using another eyrie for fledging or did not have offspring at all that year. With disappointment, I decided to look for another nest in the territory in the hope of finding that year's nesting site. As I have watched the eagles fly several times across the valley and disappear behind a small peak with nesting material or prey in their claws, I guessed that the nesting site might rather be over there. On the next sunny day, I hiked up to this peak and searched in precipitous terrain for the putative nest. But all in vain, I could not find the slightest hint of an eyrie, and concluded that the entire mountain slope was actually not an ideal place to build a golden eagle nest. Before descending into the valley, I decided first to climb much higher up the adjacent mountain to have a clear view over to the cliff where my original hide was located. The idea being that, from this position, I might be able to spy into the nest using my binoculars and inspect the back of the cave, which I had no chance of seeing from my hide at the foot of the cliff. Tired, I dropped my heavy backpack onto the ground, grabbed my binoculars, and sat down on the alpine meadows at around 2100 m above sea level. I waited for my breathing to slow down a little, positioned the binoculars, and examined the golden eagle nest at approximately 1.8 km distance. But due to thermal distortion of the air blurring my view, together with the shade in the cave, it was not clear whether the nest was empty or not. I therefore took a couple of pictures with my tele-objective over the next half an hour and thought an examination of the images later at home would be more conclusive. And indeed, after small image post-processing tweaks, such as adjusting exposure, contrast, and shadows, I could peek into the cave and examine the entire nest. And to my utmost joy there he was. A tiny, completely white eagle hatchling sat at the back of the cave, in slightly varying postures from image to image. “He’s there! Despite all my doubts, there is a golden eagle chick in that nest!” I rejoiced and started to pack my gear for the next day to visit my hide again. “I really want to see this little snow-white figure sitting all alone in the nest.” I told myself.

Although tired from the long hike on the previous day, I almost ran up the mountain to my hide, installed tripod and camera, and patiently waited for the little hatchling to crawl from the back of the cave to the rim of the nest where I could actually see him and take pictures. It was a real patience game, but after two hours I spotted a tiny white baby eagle head peeking over the edge of the nest. “There he is! Look how tiny he is.” I was deeply moved. With great satisfaction and joy, I took a couple of photos to document the moment, before continuing to observe the new golden eagle family member. To my great surprise, the fragile new inhabitant spent the entire day sitting all alone in the small cave in the middle of this large rock wall, high above the valley. “What an amazing place to grow up. But what a tough start to life.” I thought. “The essence of loneliness.”

Two weeks later, I again packed my gear and climbed up to the golden eagle nursery. Since it was already June, the vegetation had grown fast on the easterly-facing mountain slope, and the hike in the steep, grassy terrain became more and more tedious. But nothing could stop me now. I needed to know how my eaglet was doing. “I hope he is well! He must have grown quite a bit. Maybe the chick moves around a little more now.” I told myself and hoped to see more this time than just his tiny head. And I was not disappointed. The eagle chick had grown a lot and was surprisingly active. The plumage had changed from all white to a blackish-white, speckled appearance on the wings, the beak had grown larger, and the gaze out of the piercing black eyes was already impressive. From the bird’s coat, I estimated the age of the eaglet at around thirty days. But most astonishing was the behavioral change. It seemed that all his senses had been switched on. With great interest, the hatchling’s head followed every bird, butterfly, or bee that dared to come close to the nest. When a chamois mother with her young crossed the grass slope below the nest, the eaglet leaned over the rim and followed every step of his future prey with instinct-driven curiosity. And even more surprising, the young raptor constantly turned his head in my direction when I took photos. Although it was very unlikely that he could see me, as I was very well hidden in the small group of trees, the clicking noise of the camera shutter took his attention. The eaglet also changed position much more often, and even flapped his tiny wings from time to time, either to get rid of annoying parasites or to already strengthen the flight muscles. I observed my new little friend the entire day and was again impressed that he had to hold out all alone for long hours, without either of the parents paying a visit to their young son or daughter. Due to bad weather and other commitments, I only managed to visit the nest again twenty-five days later. “I wonder what has happened to the young eagle in almost three weeks? Did he make it? Is he still there? If yes, he must have grown a lot.” I said to myself during the ascent. But what I discovered in the nest after I had set up all my gear in the hide almost blew my socks off. There he was, sitting next to the nest on an elevated part of the cave. A tall, dark brown, and incredibly impressive golden eagle, featuring a large black beak framed by a bright yellow cere! Only a few small white patches on the plumage here and there reminded me that I was looking at this year’s hatchling. When the eagle lifted his feet, I was impressed by the already long, deep black talons and the beautiful bright yellow feet. “What a transition in just twenty-five days! He grew so much and completely changed his look, from a little white eaglet to this proud and tall golden eagle.” I thought. “This is a truly magnificent creature and will definitely become a king or queen of the skies one day.” I felt highly privileged to see my eagle grow up and observed the young raptor for several hours from my hide. For the first hour in the early morning, the eagle positioned his head and chest towards the sun and then turned around for another hour, exposing his back. “Ah, he regulates his body temperature by soaking up the heat of the sun with his dark plumage.” I concluded. Stock-still, the young raptor stood tall in the cave warming his body. Later in the day, I was lucky to observe more species-typical behavior of the young raptor. The eaglet changed position several times, from standing tall to sitting or leaning over the rim of the nest to follow an ungulate that passed below. He spent a lot of time in self-maintenance. He extensively preened feathers or pulled out remaining baby down, shook and ruffled his entire coat many times, or used talons to scratch his neck, head, and mouth. He yawned, adjusted small twigs in the nest with his beak, or defecated by expelling feces in a jet several meters out into the void, whilst making sure not to soil the nest. From time to time, the young eagle experienced an energy attack, and hopped back and forth within three meters of the nest and its surroundings, whilst heavily flapping his wings in a practice flight. I had the impression that there was a lot of frustration in that young bird. He looked powerful, and already had so much strength and energy. Each fiber of his body must have felt the urge to jump out of the nest for his maiden flight. I guessed the eaglet’s age at fifty-five days and estimated that he would try his first flight in around two weeks’ time. I also had the impression that the bird was hungry. Several times he looked for food debris but could not find anything, and all the food begging calls to his parents were in vain. He again stayed the entire day alone. “An incredible bird.” I thought and was convinced that the isolation during the upbringing must play an important role in forming his character. “Isolation must be part of the reason why golden eagles become such majestic and proud individuals. They get primed already at a very young age to be strong and tough.” With such ruminative thoughts in my head, I packed my gear and hiked back down to the valley, content, happy, and grateful that I could spend another day with my eagle friend.

Although the young eagle was constantly buzzing around in my thoughts during the following days, I only managed to visit the eagle’s nest three weeks later. It was towards the end of July on the first nice day just after a cold front had rushed through Switzerland, covering peaks above 2500 m in fresh snow. With a northerly wind, the hike up to the eyrie was pleasant, despite the struggle with the high summer vegetation obscuring one’s very own feet and making it difficult to find a safe foothold. As I did not want to disturb the already very observant young eagle, I decided not to approach via the ledge below the nest, where the raptor would see me, but instead to climb higher, use another rock spur to cross the eyrie on the cliff above, and then rappel down a ravine that directly led to my hide. This way, I was sure that the eagle would not notice my presence at all. “If I am lucky, I might even witness the maiden flight of the youngster. Usually golden eagles fledge during the second half of July in Switzerland, and my eaglet friend should definitely be ready for this soon.” I thought and dropped the heavy backpack in front of the small group of pines, from where I usually observed the cliff with his nest. After I had installed myself in the hide, I grabbed the binoculars and aimed with excitement at the eagle’s nest. But the young eagle was gone. Instantly a feeling of loss took over, and my excitement turned into disappointment. “Why did you do that to me? Couldn’t you have waited a little longer my friend? I will miss you. What will I do without you now? For the last three months you were a companion that gave meaning to my life. How sad that you are gone.” My thoughts became heavy. Maybe just a couple of days ago, the raptor must have taken courage, flapped his large wings with all the strength he had, then ran towards the edge of the nest and kicked himself out into the void high above the valley for the very first time. The instinctive urge to fly must have been so overwhelming that the eaglet had no chance to resist and stay longer in the nest that had been his safe place and shelter for weeks. “Did my eagle majestically fly out into the air or did he just tumble down onto the ledge below the nest in an uncontrolled way.” I asked myself.

“I will never know.” From the scientific literature I knew that fledged eaglets stay close to the nest for several weeks, where the parents continue to feed them and teach the youngsters to fly. Since the nest was in a small cave in the middle of a cliff, there was no way that the young eagle could return once he had left it. The raptor first needed to improve his flight skills. Perhaps in only two or three months would he be capable of attempting the tricky approach, and land in his former nursery ground. Therefore, I carefully inspected the surroundings of the nest with the binoculars, including the large grassy ledge below it, to see whether the eagle had found another safe place to stay. But the young raptor was nowhere to be seen. I then decided to wait for the adult eagle parents to approach their offspring somewhere nearby to deliver food, give flying lessons, or just check how their offspring was doing. My plan was to locate the youngster via the approaching parents, so that I could still visit the teenager during the following weeks to document his steps towards a mature juvenile golden eagle. I waited the entire day in my hide, but the breeding pair never approached the cliff. I spotted them on a rocky perch on the tall mountain across the valley where they seemed to hunt. I saw them circling in thermals far away, on the border of their home range, and witnessed how they crossed the valley at very high altitude in an impressive straight flight. But they never came even close to the cliff with the nest. "I really wonder where they hid the young eagle; but it looks like I will never find out." I decided to end the day and packed my gear. A little later when I stepped out of the hide onto the open alpine meadows, I noticed a white spot in the grass on the slope directly below the eyrie. "What could that be?" I thought and crossed hundred fifty meters of steep hillside to check what it was. When I reached the place where I had expected the white object, I could not believe my eyes. The white patch belonged to the base part of a beautiful feather, approximately twenty-five centimeters long. Beyond the pure white base, the color of the plume changed sharply to dark brown after around one third of its length on the terminal side. "A wing feather of my juvenile golden eagle!" I jubilated. "It looks like my friend left a message for me." "I know you. I have watched you for three months as you have watched me. Here I give you one of my feathers to thank you for your friendship and patience. Be assured that I am not gone. I have just entered a different phase of my life. Don't worry, we will see each other again." I imagined the eaglet say to me. Obviously, the juvenile eagle had lost one of his secondary wing feathers when he courageously jumped out of the nest on his maiden flight. "What a nice souvenir. Thank you my friend." I thought, picked the feather up, attached it to my backpack, and started the descent.

At the time I did not know that I would have to wait more than three months before I would finally see my eagle again. Wondering how the resident eagle pair was doing after the annual hunting season, I visited their home range again in late October. I sat on one of my favorite lookouts on top of a cliff, when suddenly golden eagles appeared around the corner gliding along the rock wall. To my great delight the familiar breeding pair was accompanied by a third, slightly smaller bird that carried the distinct white patches of a first-year juvenile golden eagle on wings and tail. "My eagle!" I thought and felt deep gratitude to see that my youngster had turned into such an impressive, skilled and healthy member of the kings of all birds. In a couple of weeks the teenager would be ready to leave his parents and start to explore vast territories throughout the Alps, and to self-reliantly lead a highly adventurous life for the next two to three years, before he would slowly come back and try to establish his own home range close to the place he was born. "Fly my eagle, fly!" I shouted out into the air and watched the young eagle turn his head and look straight into my eyes, before he took an elegant turn to the right, disappearing with his parents behind a mountain.

As a photographer I am never completely satisfied with the quality of my pictures. I always see potential for improvement. Most often it is about light. "I wished the eagle would have come an hour earlier. Its head would have been nicely lit by the sunlight and not be in the shade." Frequently it's about focus. Although I use professional gear with extremely fast autofocus capabilities, it is always a challenge to get into focus a fast flying object that suddenly and unexpectedly appears. "What a beautiful shot; but I really wish the image were crisper." I often think. Regularly it's about distance. "Wow, look at the show the eagles deliver today. They stoop, they turn at spectacular angles, they fly centimeters away from the rock wall, and the eagle over there just flipped onto its back to show its massive talons to a raven flying above him. If all the action would just happen closer to my position so that I could take a nice picture." I wish on almost every photo excursion to the mountains.

In raptor photography, light can be controlled to a certain extent, for example by anticipating where the eagle might show up with respect to the position of the sun and positioning the camera accordingly. But there is a lot of unpredictability in the process. Despite the most thorough preparations and considerations, the eagle might pay a visit an hour later when the sun is already at a different angle, approach from a totally different direction than expected or, most often, not appear at all. There are also a couple of tricks to maximize the likelihood of getting into focus a fast-moving object, like a bird in flight, to obtain a crisp image. A well-calibrated camera with an advanced continuous autofocus system, in combination with a fast telephoto prime lens, brings the image quality for sure to another level. To get the most out of such a combination it is mandatory to use it with a stable monopod or tripod. But despite the most advanced technical combinations there are factors in wildlife photography that cannot be controlled by the photographer. Thermal distortion of the air, unexpected appearance of the photographic subject, the angle of the flight trajectory with respect to the camera, or the presence of a clear or busy background; all influence focus speed and accuracy.

In contrast, the distance to the subject can always be changed by getting closer. As I know the flight patterns of golden eagles in my favorite locations quite well, I am always looking for new perspectives that would allow me to get closer to the mighty raptor and obtain higher quality photographs. Following the advice of the famous Hungarian-American war photographer Robert Capa, who once said that if your pictures aren't good enough you aren't close enough, I tirelessly analyze and search the cliffs for better positions. One winter, at a golden eagle site in the central part of Switzerland, I discovered an interesting rock ledge on the opposite south-facing wall that seemed accessible from the top of the cliff. The ledge extended for around thirty meters straight into the steepening rock wall, where it got narrower and narrower before ending in an exposed small platform on top of an overhang around eighty meters above ground. "This would be a brilliant place to photograph golden eagles. The platform is exactly located at the level of the eagle airway that the mighty raptors often take to profit from midday updrafts at the wall." I told myself. "If I could access this platform I would be very, very close to take pictures of the birds."

A couple of days later, I hiked along the high cliff edge on the other side of the valley with the aim of finding the entry to the ledge. It was a crisp January morning with a deep blue sky full of glittering ice crystals that turned the forest into a magical place of fairy tales. The fresh, deep powder from the previous night covered the landscape with a thick, fluffy white blanket which dampened all noise, so that even the shuffle of my snowshoes was almost silent. Deep tracks of chamois led from the forest to the cliff edge, where the sturdy animals must have expected the snow layer to be less deep so that they could scrabble down to the ground and satisfy their hunger with dry grass. It took me one and a half hours to reach the point on the cliff where I expected the rock spur to start. I dropped my backpack, unbuckled the snowshoes, grabbed my ice axe, and cautiously approached the edge of the wall. When I looked down the wall my heart skipped a beat. "There it is!" I exclaimed and pointed down at the rock ledge which crossed the wall horizontally, around twenty meters below me. I figured out that I had walked too far and that the entry to the rock spur was more to the right. I tramped around twenty meters to the right through the snow, and after a short search found a rock path that led through two young conifers and out into the void. "The entry to the ledge. I have found it!" I was relieved and stepped out onto the rock prominence. Meter by meter, I shoveled the snow away and followed the narrowing rock terrace into the wall to the very end. More crawling than walking, I reached the end of the ledge and sat down on the small, exposed pedestal that measured around three square meters. "I am in the midst of the rock wall. What a breathtaking view! It's an exceptional place to photograph golden eagles!" I said to myself. "This will be my new office for the winter." I laughed and went all the way back to get my gear.

During the following two months, I regularly spent time in my new office to photograph golden eagles. As anticipated, the ledge was precisely located on the level at which the kings of the sky, with the help of the south-facing updrafts, soared across the cliff to reach an amphitheater-like rock formation where, in winter, strong circular thermals develop. At first, the resident eagle pair was obviously surprised to find a visitor on their flightpath and started to carefully investigate the change. They interrupted their straight, gliding flight along the wall as soon as they saw me and circled at a safe distance around the platform, inspecting me. But after several visits, they had learned that there was no danger, and the huge eagles whizzed at full speed past me at just five meters distance, so that I not only heard the loud swoosh of their wings cutting through the air, but even felt a breeze on my face. Over several sessions I obtained great close-up photographs of the raptors buzzing by, which fulfilled me with great joy and satisfaction. Often, I was really happy with the yield of new photographs, after spending another sunny winter day high up on my exposed platform in the rock wall. Happy and content, I would then walk back down to the valley and murmur: "I am so glad I could spend another day in the office with my feathered friends."

One summer, I spent a lot of time in the mountains where I photographed B.A.S.E. jumpers on their spectacular proximity flights, along cliffs and through deep ravines. Using my experience as a rock climber, I researched locations and found sites in rock walls, on exposed spires, or in deep canyons, where the human birds in their wingsuits whizzed by, close enough to obtain mind-blowing action photographs. During this summer, taking pictures of B.A.S.E. jumpers was my personal action sport, without risking my life like my photographic subjects. Dangling from a rope on a cliff two hundred meters above ground with my camera, my adrenalin levels spiked every time I heard the characteristic swoosh of an approaching human flying squirrel. In these moments, I knew that I had five seconds to find the B.A.S.E. jumper in my tele-objective, focus, follow the brave action addict through the air, and get a series of pictures of the flying object that passed my position just a few meters away at an insane speed of one hundred fifty to two hundred kilometers per hour. Aware of the risks of this demanding sport, costing several lives every season in the Swiss mountains, I was always relieved to hear the pop of the parachute opening, meaning that my unaware photographic subject likely will touch ground in a controlled manner and soon be back on earth safe and sound.

As I prefer to photograph life as it happens, I don't like staged photography where the action hero and the photographer precisely orchestrate location, shooting angle, light, or color of the gear and outfit. Therefore, I had to patiently wait for the action in a very similar way to the wildlife photography that I know so very well. To my delight, I realized that the playground of B.A.S.E. jumpers in Switzerland was in areas of the mountains where untouched nature with high-quality habitats provide a home for wild animals and many species of birds of prey. So, whenever there was no human action in the air, I took pictures of the local flora and fauna. At one site, located on a tall cliff above a canyon, I soon became familiar with the local kestrel pair that was highly busy keeping under control this year's fledgling, which had just taken its maiden flight. Sometimes, the youngster landed on a small ledge just a few meters away from me, where it started with endless food-begging calls that echoed for minutes from the surrounding rock walls, until one of the parents finally showed up delivering a mouse or a lizard. From observing these birds, I learned that kestrel parents are highly territorial and protective of their offspring. Whenever they discovered a potentially dangerous bird approaching their cliff, they started a brave raid against the intruder who sometimes surpassed the small falcon's size by more than ten times. With loud shrieks the rufous-colored raptors pounced on everything that moved. Eurasian buzzards, ravens, peregrines, or paragliders, no one was spared! Usually these territorial kestrel attacks were so hefty and annoying that the intruder quickly decided to leave these cranky locals in peace and navigate out of their airspace as fast as possible.

But one day, something different happened. It was on a peaceful morning when the kestrels suddenly raised an alarm and started their interception flight with great agitation. Being used to the scenario, I thought that another raven or buzzard was on its way towards the cliff. But after a minute, the shrieks of the kestrels were accompanied by an entire orchestra of alarm calls. The complete forest was on alert. Jays, craws, choughs, and the local peregrine, all sounded alarms or changed their flight patterns. Even the ever-busy squirrel in the pine tree next to me stopped chewing on its cone and froze in total immobility. When I looked over to the other side of the canyon, I realized that the small group of red deer, that had been quietly grazing just a minute ago, had disappeared. "What is going on here?" I thought and scanned the surroundings with tension and excitement. For a long minute nothing happened. But then, suddenly, not only the animals' but also my adrenalin level spiked. A huge golden eagle flew from the left, around the edge of the cliff, and cruised like a B52 just five meters below my position along the rock wall. Already from around twenty meters distance, I heard the swooshing sound of the immense wings, and when the eagle passed right under me, I could look straight into the king's hazel colored eyes. Only when the large bird disappeared to the right, flying deep into the dark canyon, did I wake up, grab my camera, and take a series of photographs. As B.A.S.E. jumpers often dive straight through this canyon, I just hoped that there wouldn't be a collision between the two. It most likely would have ended fatally for both players. But soon after, the golden eagle came out of the canyon and majestically glided along the opposite wall of the gorge. It was at this moment when the kestrels attacked. With shrill war cries, they took all their courage and stooped straight at the immense eagle, only to change trajectory a meter before impact. The eagle at first only turned its head towards the annoying raiders and continued its flight. But when the falcons untiringly continued their attacks over the following minutes, the eagle got more and more uneasy. It sometimes flipped onto its back and tried to snatch one of the small agile raptors, but never managed. The size difference between the birds was extreme. The eagle must have felt the same irritation as when we try to get rid of an annoying wasp or fly that buzzes around our head. To my delight I could not only witness an amazing wildlife scene, but also managed to take many photographs of the quarrel. After around five more minutes, the golden eagle got so frustrated by the never-ending strikes, that it decided enough is enough. In a sharp turn it left the wall of the cliff and glided straight out into the open sky high above the forest. Probably realizing that they had won the fight, the kestrels terminated the chase, screeched a few more times, and victoriously returned to their nest. I have no doubt that the courageous little falcons have earned huge respect from the king of the air who probably will think twice before flying through this canyon again.

THE EAGLE HAS LANDED.

NEIL ARMSTRONG \
1930 — 2012

This book would not have been possible without the help, patience, and support of family and friends, all of whom I am greatly thankful. A special thanks goes to J. Copley, S. Saxena, and B. Miranda for correcting the manuscript, and to A.R. Garner for sharing ideas on content and design. I am, above all, grateful to my mother who never stopped believing in me and who always lovingly encouraged me in all endeavors, no matter how adventurous as they might have been. I am similarly thankful to my father who aroused my great passion for the mountains that has persisted through my entire life already from a very young age, and that has enabled me to visit golden eagles in their rugged mountain homes. Lastly, I would like to thank all the fantastic people that I encountered on my travels around the globe and that share my passion for nature and wildlife. We all must work together and continue to undertake the greatest efforts to protect it.

\01\ PLINIUS SECUNDUS, G., J. BOSTOCK, and H.T. RILEY, *The Natural History of Pliny. Vol. 1 Book 10: The Natural History of Birds. Chapter 3: The Different Kind of Eagles.* 1855, London, UK: Henry G. Bohn.

\02\ LERNER, H.R.L. and D.P. MINDELL, *Phylogeny of Eagles, Old World Vultures, and Other Accipitridae Based on Nuclear and Mitochondrial DNA.* Molecular Phylogenetics and Evolution, 2005. 37(2): p. 327–46.

\03\ WINKLER, D.W., S.M. BILLERMAN, and I.J. LOVETTE, *Hawks, Eagles, and Kites (Accipitridae), Version 1.* Birds of the World, ed. S.M. BILLERMAN, et al. 2020, Ithaca, New York, USA: Cornell Lab of Ornithology.

\04\ AGUIAR-SILVA, F.H., T.M. SANAIOTTI, and B.B. LUZ, *Food Habits of the Harpy Eagle, a Top Predator from the Amazonian Rainforest Canopy.* Journal of Raptor Research, 2014. 48(1): p. 24–35.

\05\ CHIWESHE, N., *Black Eagles and Hyraxes – The two Flagship Species in the Conservation of Wildlife in the Matobo Hills, Zimbabwe.* Ostrich, 2007. 78(2).

\06\ GARGETT, V., *The Black Eagle: A Study.* 1990, Randburg, South Africa: Acorn Books and Russel Friedman Books.

\07\ SIMMONS, R., *Offspring Quality and the Evolution of Cainism.* Ibis, 1988. 130(3): p. 339–357.

\08\ DALAL, R., *Hinduism: An Alphabetical Guide.* 2011, London, UK: Penguin Books.

\09\ WILLIAMS, G.M., *Handbook of Hindu Mythology.* 2008, Oxford, UK: Oxford University Press.

\10\ GOPINATHA RAO, T.A., *Elements of Hindu Iconography.* 1993, Delhi, India: Motilal Banarsidass.

\11\ LARRINGTON, C., *The Poetic Edda.* 2019, Oxford, UK: Oxford University Press.

\12\ STURLUSON, S. and A. FAULKES, *Edda.* 1995, London, UK: Everyman Paperbacks.

\13\ LINDOW, J., *Norse Mythology: A Guide to the Gods, Heroes, Rituals, and Beliefs.* 2002, Oxford, UK: Oxford University Press.

\14\ ROSCHER, W.H., *Ausführliches Lexikon der Griechischen und Römischen Mythologie.* 1924–1937, Leipzig, Germany: B. G. Teubner.

\15\ COTTERELL, A., *The Pimlico Dictionary of Classical Mythologies.* 2000, London, UK: Pimlico Random House.

\16\ PAYMENT, S., *Greek Mythology.* 2006, New York, USA: Rosen Publishing Group.

\17\ LORBER, C.C., *Coins of the Ptolemaic Empire: Ptolemy I-Ptolemy IV.* 2018, New York, USA: American Numismatic Society.

\18\ ARISTOTLE, *The Works of Aristotle. Translated into English Under the Editorship of J. A. Smith and W. D. Ross. Volume IV Historia Animalium by D'Arcy Wentworth Thompson.* 1910, Oxford, UK: Clarendon Press.

\19\ RAHBERG, S.D., *Enduring Ministry: Toward a Lifetime of Christian Leadership.* 2017, Collegeville, Minnesota, USA: Liturgical Press.

\20\ HEJDUK, J.D., *The God of Rome: Jupiter in Augustan Poetry.* 2020, Oxford, UK: Oxford University Press.

\21\ SMITH, W., W. WAYTE, and G.E. MARINDIN, *A Dictionary of Greek and Roman Antiquities.* 1890, London, UK: John Murray.

\22\ KHARIOUZOV, A., *Prodigien in der Römischen Königszeit: Eine Motivgeschichtliche und Narratologische Analyse im 1. Buch des Livius.* Klassische Philologie. Vol. 5. 2014, Berlin, Germany: Frank & Timme.

\23\ LOVANO, M., *All Things Julius Caesar: An Encyclopedia of Caesar's World and Legacy.* 2017, Santa Barbara, California, USA: Greenwood.

\24\ TÖPFER, K., SIGNA MILITARIA: *Die Römischen Feldzeichen in der Republik und im Prinzipat.* 2011, Mainz, Germany: Verlag der Römisch-Germanischen Zentralmuseums.

\25\ ANTIOCH, H.O., *History of the Roman Empire : From the Death of Marcus Aurelius to the Accession of Gordian III. Translated from the Greek by Edwards C. Echols.* 1961, Berkeley, Los Angeles, California, USA: University of California Press.

\26\ GIESE, M., *Der Adler als Kaiserliches Symbol in Staufischer Zeit.* Staufisches Kaisertum im 12. Jahrhundert, eds. Stefan Burkhardt, Thomas Metz, Bernd Schneidmüller, Stefan Weinfurter. 2010, Regensburg, Germany: Schnell & Steiner. 323–360.

\27\ RADER, O.B., *Friedrich der Zweite: der Sizilianer auf dem Kaiserthron.* 2010, Munich, Germany: C. H. Beck.

\28\ MENZEL, W., *Christliche Symbolik.* 1854, Regensburg, Germany: G. Joseph Manz.

\29\ MÂLE, E., *Religious Art in France of the Thirteenth Century.* 2000, Mineola, New York, USA: Dover Publications.

\30\ FLORES, N.C., *Animals in the Middle Ages.* 1996, Abingdon, New York, USA: Routledge.

\31\ RIPA, C. and E.A. MASER, *Baroque and Rococo Pictorial Imagery: The 1758–60 Hertel Edition of Ripa's 'Iconologia' with 200 Engraved Illustrations.* 1991, New York, USA: Dover Publications.

\32\ LAKE-THOM, B., *Spirits of the Earth: A Guide to Native American Nature Symbols, Stories, and Ceremonies.* 1997, New York, USA: Plume.

\33\ LARDNER, D.C. AND J. CARMODY, *Native American Religions: An Introduction.* 1993, New York, USA: Paulist Press.

\34\ TAYLOR, C.F., *Native American Myths and Legends.* 1996, Vancouver, Canada: Cavendish Books.

\35\ LYNCH, P.A. AND J. ROBERTS, *Native American Mythology A to Z.* 2010, New York, USA: Chelsea House Publishers.

\36\ OWUSU, H., *Symbols of Native America.* 1999, New York, USA: Sterling Publishing Co. Inc.

\37\ LAUBIN, R. and G. LAUBIN, *Indian Dances of North America: Their Importance to Indian Life.* 1989, Norman, Oklahoma, USA: University of Oklahoma Press.

\38\ HODGE, F.W., *Handbook of American Indians North of Mexico.* Vol. 3. 1912, Washington, DC, USA: Washington Government Printing Office.

\39\ HARDIN, B.E., *The Plains Warbonnet: Its Story and Construction.* 2013, Pottsboro, Texas, USA: Crazy Crow Trading Post.

\40\ WISCH, R.F., *Detailed Discussion of the Bald and Golden Eagle Protection Act.* Animal Law Legal and Historical Center, 2002.

\41\ HULL, K. and R. FERGUS, *Eagles in Mesoamerican Thought and Mythology.* Reitaku Review, 2009. 15.

\42\ RICE, P.M., *Maya Calendar Origins.* 2007, Austin, Texas, USA: University of Texas Press.

\43\ AGUILAR-MORENO, M., *Handbook to Life in the Aztec World.* 2007, Oxford, UK: Oxford University Press.

\44\ BIERHORST, J., *A Nahuatl-English Dictionary and Concordance to the Cantares Mexicanos: With an Analytic Transcription and Grammatical Notes.* 1995, Stanford, California, USA: Stanford University Press.

\ 45 \ SULLIVAN, L.E., *Native Religions and Cultures of Central and South America: Anthropology of the Sacred.* 2002, New York, USA: Continuum International Publishing Group.

\ 46 \ GRUBE, N. and T. BÜRGER, *Der Dresdner Maya-Kalender: Der vollständige Codex.* 2012, Freiburg im Breisgau, Basel, Vienna: Herder Verlag.

\ 47 \ MEYER, F.S., *Handbook of Ornament: A Grammar of Art Industrial and Architectural Designing.* 1957, New York, USA: Dover Publications.

\ 48 \ GARDNER, H., *National Flag Design. Studies of Primary Design Elements, Colors, Symbols.* 2013, Pittsburgh, Pennsylvania, USA: Dorrance Publishing Company Inc.

\ 49 \ WHITMORE, W.H., *The Elements of Heraldry: Containing an Explanation of the Principles of the Science and a Glossary of the Technical Terms Employed. With an Essay upon the Use of Coat-Armor in the United States.* 1970, New York, USA: Weathervane Books.

\ 50 \ PARISIENSIS, M., *Chronica Majora Vol. 4 A.D. 1240 to A.D. 1247,* ed. H.R. Luard. 1877, London, UK: Longman & Co.

\ 51 \ RADY, M., *Habsburg Empire: A Very Short Introduction.* 2017, Oxford, UK: Oxford University Press.

\ 52 \ KAZHDAN, A.P., A. CUTLER, and A.-M. TALBOT, *The Oxford Dictionary of Byzantium.* 1991, Oxford, UK: Oxford University Press.

\ 53 \ SOLOVIEV, A.V., *Seminarium Kondakovianum. Vol. 7. Les Emblèmes Héraldiques de Byzance et les Slaves.* 1935, Prague, Czech Republic: Institut Kondakov.

\ 54 \ YILMAZ, S., *Putin Era Russia's Foreign Policy and Security Doctrines.* 2020, Ankara, Turkey: Ilbilge Yayincilik Egitim.

\ 55 \ GROSBY, S.E. and A.S. LEOUSSI, *Nationalism and Ethnosymbolism – History, Culture and Ethnicity in the Formation of Nations.* 2007, Edinburgh, UK: Edinburgh University Press.

\ 56 \ WRÓBEL, P., *Historical Dictionary of Poland 1945–1996.* 2014, London, UK: Routledge.

\ 57 \ WISE, T., et al., *Flags of the Napoleonic Wars (1) – Colours, Standards and Guidons of France and her Allies.* 1978, London, UK: Osprey Publishing Ltd.

\ 58 \ FRASER, E., *The War Drama of the Eagles – Napoleon's Standard-Bearers on the Battlefield in Victory and Defeat from Austerlitz to Waterloo.* 2011, Auckland, New Zealand: Pickle Partners Publishing.

\ 59 \ MUCHA, L. and W.G. CRAMPTON, *Webster's Concise Encyclopedia of Flags & Coats of Arms.* 1987, New York, USA: Crescent Books.

\ 60 \ PATTERSON, R.S. and R. DOUGALL, *The Eagle and the Shield : A History of the Great Seal of the United States.* 1976, Washington, DC, USA: Department of State, Office of the Historian, Bureau of Public Affairs.

\ 61 \ COMMUNICATION, O.O.P., *The Great Seal of the United States.* 2003.

\ 62 \ HATHAWAY, J., *A Tale of Two Factions – Myth, Memory, and Identity in Ottoman Egypt and Yemen.* 2003.

\ 63 \ TOFT, R., *National Birds of the World.* 2014, London, UK: Bloomsbury.

\ 64 \ MINAHAN, J., *The Complete Guide to National Symbols and Emblems.* 2010, Santa Barbara, California, USA: Greenwood Press.

\ 65 \ MÜHLEMANN, L., *Wappen und Fahnen der Schweiz, 700 Jahre Confoederatio Helvetica: Offizieller Wappen- und Fahnen-Bildband zur 700-Jahrfeier der Eidgenossenschaft.* 1996, Lengnau, Switzerland: Bühler.

\ 66 \ GEOINFORMATION, I.O.C.A., *Atlas der Schweiz.* ETH Zurich, 2020.

\ 67 \ BIRDLIFE-INTERNATIONAL, *Aquila chrysaetos. The IUCN Red List of Threatened Species.* 2016. e.T22696060A93541662.

\ 68 \ WATSON, J., K. BROCKIE, and D. WATSON, *The Golden Eagle.* 2011, New Haven, Connecticut, USA: Yale University Press.

\ 69 \ MCINTYRE, C.L., D.C. DOUGLAS, and M.W. COLLOPY, *Movements of Golden Eagles (Aquila chrysaetos) from Interior Alaska During Their First Year of Independence.* The Auk, 2008. 125(1): p. 214–224.

\ 70 \ SHERRINGTON, P., *The Road to the Place of the Eagles.* BirdWatch Canada, 2008. 43.

\ 71 \ LINNAEUS, C., *Systema Naturae per Regna Tria Naturae: Secundum Classes, Ordines, Genera, Species, cum Characteribus, Differentiis, Synonymis, Locis.* 1758, Holmiae, Sweden: Laurentius Salvius.

\ 72 \ BRISSON, M.J., *Ornithologia sive synopsis methodica sistens avium divisionem in ordines, sectiones, genera, species, ipsarumque varietates.* 1760, Paris, France: Bauche, J.B.

\ 73 \ JOBLING, J.A., *The Helm Dictionary of Scientific Bird Names : From Aalge to Zusii.* 2010, London, UK: Christopher Helm, A&C Black Publishers Ltd.

\ 74 \ KOCHERT, M.N., et al., *Golden Eagle (Aquila chrysaetos). Birds of the World, ed. S.M. Billerman.* 2020, Ithaca, New York, USA: Cornell Lab of Ornithology.

\ 75 \ FERGUSON-LEES, J. and D.A. CHRISTIE, *Raptors of the World.* 2001, London, UK: Christopher Helm, A & C Black Publishers Ltd.

\ 76 \ BORTOLOTTI, G.R., *Age and Sex Size Variation in Golden Eagles.* Journal of Field Ornithology, 1984. 55(1): p. 54–66.

\ 77 \ FORSMAN, D., *Flight Identification of Raptors of Europe, North Africa and the Middle East.* 2016, London, UK: Christopher Helm, Bloomsbury Publishing Plc.

\ 78 \ LIGUORI, J., *How to Age Golden Eagles.* Birding, 2004. 6.

\ 79 \ JOLLIE, M., *Plumage Changes in the Golden Eagle.* The Auk, 1947. 64.

\ 80 \ LOCKHART, J.M., *Raptor: A Journey through Birds.* 2017, Chicago, USA: The University of Chicago Press.

\ 81 \ ASPINALL, V., M. CAPPELLO, and C. PHILLIPS, *Introduction to Veterinary Anatomy and Physiology Textbook.* 2015, Amsterdam, Netherlands: Elsevier Ltd.

\ 82 \ GONZALEZ-MARTIN-MORO, J., J.L. HERNANDES-VERDEJO, and A. CLEMENT-CORRALL, *The Visual System of Diurnal Raptors: Updated Review.* Archivos de la Sociedad Espanola de Oftalmologia, 2017. 92(5): p. 225–232.

\ 83 \ MITKUS, M., et al., *Raptor Vision.* Oxford Research Encyclopedia, Neuroscience. 2020, Oxford, UK: Oxford University Press.

\ 84 \ LAMB, T.D., S.P. COLLIN, and E.N. PUGH, JR., *Evolution of the Vertebrate Eye: Opsins, Photoreceptors, Retina and Eye Cup.* Nature Reviews Neuroscience, 2007. 8(12): p. 960–76.

\ 85 \ JONAS, J.B., U. SCHNEIDER, and G.O.H. NAUMANN, *Count and Density of Human Retinal Photoreceptors.* Graefe's Archive for Clinical and Experimental Ophthalmology, 1992. 230(6): p. 505–10.

\ 86 \ LAND, M.F. and D.-E. NILSSON, *Animal Eyes.* 2002, Oxford, UK: Oxford University Press.

\ 87 \ BRINGMANN, A., *Structure and Function of the Bird Fovea.* Anatomia Histologia Embryologia, 2019. 48(3): p. 177–200.

\ 88 \ TUCKER, V.A., *The Deep Fovea, Sideways Vision and Spiral Flight Paths in Raptors.* The Journal of Experimental Biology, 2000. 203(Pt 24): p. 3745–54.

\ 89 \ HALLER, H., *Der Steinadler in Graubünden: Langfristige Untersuchungen zur Populationsökologie von Aquila chrysaetos im Zentrum der Alpen.* Der Ornithologische Beobachter Beiheft 9, 1996.

\ 90 \ KNAUS, P., et al., *Schweizer Brutvogelatlas 2013–2016: Verbreitung und Bestandesentwicklung der Vögel in der Schweiz und im Fürstentum Liechtenstein.* 2018, Sempach, Switzerland: Schweizerische Vogelwarte.

\ 91 \ ZECHNER, L., *Siedlungsdichte und Reproduktion des Steinadlers (Aquila chrysaetos) in den Südlichen Niederen Tauern (Steiermark).* Abhandlungen der Zoologisch-Botanischen Gesellschaft in Österreich, 1996. 29: p. 123–139.

\ 92 \ CLOUET, M., C. BARRAU, and J.-L. GOAR, *The Golden Eagle (Aquila chrysaetos) in the Balé Mountains, Ethiopia.* Journal of Raptor Research, 1999. 33(2): p. 102–109.

\ 93 \ COLLOPY, M.W. and T.C. EDWARDS, *Territory Size, Activity Budget, and Role of Undulating Flight in Nesting Golden Eagles* Journal of Field Ornithology, 1989. 60(1): p. 43–51.

\ 94 \ BRENDEL, U., *Der Steinadler in den Alpen – Lebensweise und Schutz.* Jahrbuch des Vereins zum Schutz der Bergwelt. Vol. 68/69. 2003/2004, Munich, Germany.

\ 95 \ COLLOPY, M.W., *Foraging Behavior and Success of Golden Eagles.* The Auk, 1983. 100(3): p. 747–749.

\ 96 \ HARMATA, A.R., *What is the Function of Undulating Flight Display in Golden Eagles.* Raptor Research, 1982. 16(4): p. 103–109.

\ 97 \ ELLIS, D.H., *Development of Behavior in the Golden Eagle.* Wildlife Monographs. Vol. 70. 1979: The Wildlife Society.

\ 98 \ KELLER, V., et al., *Rote Liste Brutvögel.* Gefährdete Arten der Schweiz, Stand 2010. 2010, Bern, Switzerland: Bundesamt für Umwelt BAFU, und Schweizerische Vogelwarte, Sempach.

\ 99 \ MACDONALD, G., *The Sanity of William Blake.* 1966, New York, USA: Haskell House.

\ 100 \ KELLER, W., *Am Kaminfeuer der Tessiner. Sagen und Volksmärchen.* 1940, Zurich, Switzerland: M. S. Metz.

\ 101 \ LEÓN-PORTILLA, M., *Fifteen Poets of the Aztec World.* 2000, Norman, Oklahoma, USA: University of Oklahoma Press.

\ 102 \ LAKE, M.G., *Native Healer: Initiation into an Ancient Art.* 1991, Wheaton, Illinois, USA: Quest Books.

\ 103 \ MICHEL, H., *Ein Kratten voll Lauterbrunnen Sagen.* 1937, Interlaken, Switzerland: O. Schläfli Buch- und Kunstdr. A.-G.

\ 104 \ PAINE, T.O., *Selections from the Poems of Timothy Otis Paine.* 1897, New York, USA, London, UK: G. P. Putnam's Sons.

\ 105 \ JECKLIN, D., *Volksthümliches aus Graubünden.* 1874, Zurich, Switzerland: Orell Füssli & Co.

\ 106 \ STREET, A.B., *The Poems of Alfred B. Street.* 1847, New York, USA: Clark & Austin.

\ 107 \ ADAMS, R.C. and D.L. NICHOLS, *Legends of the Delaware Indians and Picture Writing.* 2000, Syracuse, New York, USA: Syracuse University Press.

\ 108 \ VICTOR, F.F., *Poems of Sentiment and Imagination, with Dramatic and Descriptive Pieces. By Frances A. and Metta V. Fuller.* 1851, New York, USA: A.S. Barnes & Co.

\ 109 \ FIRE, J. and R. ERDOES, *Lame Deer. Seeker of visions. The Life of a Sioux Medicine Man.* 1972, New York, USA: Simon & Schuster.

\ 110 \ LESSING, G.E., *Gotthold Ephraim Lessings Fabeln: Drei Bücher.* 1815, Milan, Italy: K.K. Regierungs-Buchdruckerei.

\ 111 \ SULLIVAN, T.D. and T.J. KNAB, *A scattering of jades: stories, poems, and prayers of the Aztecs.* 2003, Tucson, Arizona, USA: University of Arizona Press.

\ 112 \ STURLUSON, S. and A.G. BRODEUR, *The Prose Edda.* 1916, New York, USA: The American-Scandinavian Foundation.

\ 113 \ AESCHYLUS, *Four Plays of Aeschylus.* Rendered into English Verse by G. M. Cookson. 1922, Oxford, UK: Basil Blackwell.

\ 114 \ PERCIVAL, J.G., *The Dream of a Day, and Other Poems.* 1843, New Haven, Connecticut, USA: S. Babcock.

\ 115 \ SHELLEY, P.B., *The revolt of Islam; a poem.* 1818, London, UK: Printed for C. and J. Ollier, by B. McMillan.

\ 116 \ TENNYSON, A.L., *The poems of Lord Tennyson 1830–1858.* 1902, London, UK: Grant Richards.

IMPRINT

The Deutsche Nationalbibliothek lists this publication in the Deutsche Nationalbibliografie; detailed bibliographic data are available on the Internet at http://dnb.dnb.de.

ISBN 978-3-7165-1866-3

www.benteli.ch

1st edition 2022

Book concept, text, and photographs
Heinz M. Gut
Graphic concept and layout Eva Finkbeiner
Reproduction Bild1Druck GmbH, Berlin

PERMISSIONS

The Warrior and the Eagle: in R.C. Adams, Legends of the Delaware Indians and Picture Writing, Syracuse University Press, 1997, Syracuse, New York, USA. Reprinted with the kind permission of Syracuse University Press.

Axayacatl's army marches to Purepecha: in M. Leon-Portilla, Fifteen Poets of the Aztec World, University of Oklahoma Press, 2000, Norman, Oklahoma, USA. Reprinted with the kind permission of University of Oklahoma Press.

Soaring like an eagle: From A Scattering of Jades: Stories, Poems, and Prayers of the Aztecs by Timothy J. Knab. © 2003 The Arizona Board of Regents. Reprinted by permission of the University of Arizona Press.